Applied Mathematical Sciences

EDITORIAL STATEMENT

The mathematization of all sciences, the fading of traditional scientific boundaries, the impact of computer technology, the growing importance of mathematical-computer modelling and the necessity of scientific planning all create the need both in education and research for books that are introductory to and abreast of these developments.

The purpose of this series is to provide such books, suitable for the user of mathematics, the mathematician interested in applications, and the student scientist. In particular, this series will provide an outlet for material less formally presented and more anticipatory of needs than finished texts or monographs, yet of immediate interest because of the novelty of its treatment of an application or of mathematics being applied or lying close to applications.

The aim of the series is, through rapid publication in an attractive but inexpensive format, to make material of current interest widely accessible. This implies the absence of excessive generality and abstraction, and unrealistic idealization, but with quality of exposition as a goal.

Many of the books will originate out of and will stimulate the development of new undergraduate and graduate courses in the applications of mathematics. Some of the books will present introductions to new areas of research, new applications and act as signposts for new directions in the mathematical sciences. This series will often serve as an intermediate stage of the publication of material which, through exposure here, will be further developed and refined. These will appear in conventional format and in hard cover.

MANUSCRIPTS

The Editors welcome all inquiries regarding the submission of manuscripts for the series. Final preparation of all manuscripts will take place in the editorial offices of the series in the Division of Applied Mathematics, Brown University, Providence, Rhode Island.

SPRINGER-VERLAG NEW YORK INC., 175 Fifth Avenue, New York, N. Y. 10010

Printed in U.S.A.

Applied Mathematical Sciences | Volume 43

Hilary Ockendon
Alan B. Tayler

Inviscid Fluid Flows

With 45 Illustrations

Springer-Verlag
New York Heidelberg Berlin

Hilary Ockendon
Somerville College
Oxford, England OX2 6HD

Alan B. Tayler
Mathematical Institute
24-29 St. Giles
Oxford, England OX1 3LB

AMS Subject Classification: 76-01, 76BXX

Library of Congress Cataloging in Publication Data
Ockendon, Hilary
Inviscid fluid flows.
(Applied mathematical sciences; v.43)
Bibliography: p.
Includes index.
1. Fluid dynamics. I. Tayler, Alan B. II. Title.
III. Series: Applied mathematical sciences (Springer-Verlag New York, Inc.); v. 43.
QA1.A647 vol. 43 510s [532'.05] 83-378
[QA911]

Printed and bound by R.R. Donnelley & Sons, Harrisonburg, VA.
Printed in the United States of America.

9 8 7 6 5 4 3 2 1

ISBN 0-387-**90824**-2 Springer-Verlag New York Heidelberg Berlin
ISBN 3-540-**90824**-2 Springer-Verlag Berlin Heidelberg New York

Preface

Applied Mathematics is the art of constructing mathematical models of observed phenomena so that both qualitative and quantitative results can be predicted by the use of analytical and numerical methods. Theoretical Mechanics is concerned with the study of those phenomena which can be observed in everyday life in the physical world around us. It is often characterised by the macroscopic approach which allows the concept of an element or particle of material, small compared to the dimensions of the phenomena being modelled, yet large compared to the molecular size of the material. Then atomic and molecular phenomena appear only as quantities averaged over many molecules. It is therefore natural that the mathematical models derived are in terms of functions which are continuous and well behaved, and that the analytical and numerical methods required for their development are strongly dependent on the theory of partial and ordinary differential equations. Much pure research in Mathematics has been stimulated by the need to develop models of real situations, and experimental observations have often led to important conjectures and theorems in Analysis. It is therefore important to present a careful account of both the physical or experimental observations and the mathematical analysis used.

The authors believe that Fluid Mechanics offers a rich field for illustrating the art of mathematical modelling, the power of mathematical analysis and the stimulus of applications to readily observed phenomena. Mathematical models in Fluid Mechanics are frequently nonlinear and their solution is challenging. The material selected for discussion in these notes has been chosen both for its mathematical interest and its physical relevance. It is hoped that the topics chosen provide instructive examples of a mathematical investigation of a real problem and that, by analogy, the reader may be equipped to tackle an apparently unrelated problem in a different field. We have restricted the discussion in these notes to the flow of *inviscid* (not viscous) fluids. Even with this restriction it is possible to describe striking phenomena such as tidal waves and sonic booms.

For the past ten years, the notes have been used at Oxford University in association with a course given to final year undergraduates and first year graduates in Mathematics. It is assumed that the reader has covered standard elementary material on inviscid incompressible hydrodynamics and has had an introduction to partial differential equations and wave motion. Sections which are asterisked are included for completeness but may be omitted if a shorter course is required. Most students taking the Oxford course would also in the same year take a corresponding course on viscous flow in which the ideas of boundary layer theory would be discussed in detail, together with the relevance of the inviscid model. A suitable reference for this material is Batchelor [1]. Exercises are given at the end of each chapter, many of which are modified versions of examination questions in the Oxford Final Honour School of Mathematics.

We are both very grateful for the valuable advice and criticism given by Dr. J. R. Ockendon in the preparation of these notes.

Contents

Chapter I
Mathematical Models Of Fluid Flows

1. ELEMENTARY FLUID DYNAMICS

Basic Ideas

In this book we are concerned with mathematical solutions to the equations of fluid flow. We shall consider almost exclusively situations in which viscosity can be neglected but will deal with both incompressible and compressible fluids. In this chapter we derive the equations of flow. We shall do this briefly so that the equations are available and the underlying assumptions made clear. Further details of the derivations and justifications of these assumptions are available in many books on Fluid Dynamics; reference may be made to Batchelor [1], for example.

A fluid is regarded as a continuum; that is, the fluid particles have the same topological relationship with each other at all times. The state of the fluid may be described in terms of its velocity q, pressure p, density ρ and temperature T. If the independent variables are x and t, where x is a three dimensional vector with components (x_1,x_2,x_3) referred to inertial cartesian axes and t is time, then we have an *Eulerian* description of the flow. An alternative description in which attention is focused on a fluid particle is obtained by using a,t as independent variables, where a is the initial position of the particle. This is a *Lagrangian* description. A particle path $x = x(a,t)$ is obtained by integrating $\frac{dx}{dt} = q$ with $x = a$ at $t = 0$, and this relation may be used to change from Eulerian to Lagrangian variables. The two descriptions are equivalent and for most problems the Eulerian is found to be more useful. In the Eulerian description it is important to distinguish between differentiation following a fluid particle which is denoted by $\frac{d}{dt}$, and differentiation at a fixed point which is denoted by $\frac{\partial}{\partial t}$. If $c(x,t)$ is any differentiable function of x and t then

$$\frac{dc}{dt} = \frac{\partial c}{\partial t} + (q\cdot\nabla)c, \tag{1.1}$$

where ∇ is the gradient operator with respect to x components. Then $\frac{dc}{dt}$ is called the *convective derivative* of c, and $(q\cdot\nabla)c$ is the convective term which takes account of the motion of the fluid particles.

We have already assumed that the fluid is a continuum, and this implies that the transformation $a \to x$ is a continuous transformation of Euclidean space onto itself with time as the transformation parameter. This transformation must be one to one and have an inverse. Another consequence of the continuum hypothesis is that fluid particles which are on the boundary of a fluid region at any time must always remain on the boundary. Hence, if the boundary of the fluid is given by $f(x,t) = 0$, then on that boundary

$$\frac{df}{dt} = 0 = \frac{\partial f}{\partial t} + q\cdot\nabla f. \tag{1.2}$$

For a fixed solid boundary $\frac{\partial f}{\partial t} = 0$ and the boundary condition is

$$q\cdot\nabla f = 0 \quad \text{or} \quad q_n = 0, \tag{1.3}$$

where q_n is the component of q normal to the surface $f = 0$. If a solid boundary is moving then (1.2) implies that the relative normal velocity is zero.

The transformation $a \to x$ will have a Jacobian $J(x,t) = \frac{\partial(x_1,x_2,x_3)}{\partial(a_1,a_2,a_3)}$ which represents the physical dilatation (i.e., expansion or contraction) of a small element following the fluid. Since the transformation is invertible and continuous, J will be bounded and non-zero and we can find the convective derivative of J:

$$\begin{aligned}\frac{dJ}{dt} &= \frac{\partial(\dot{x}_1,x_2,x_3)}{\partial(a_1,a_2,a_3)} + \frac{\partial(x_1,\dot{x}_2,x_3)}{\partial(a_1,a_2,a_3)} + \frac{\partial(x_1,x_2,\dot{x}_3)}{\partial(a_1,a_2,a_3)}\\ &= \frac{\partial(q_1,x_2,x_3)}{\partial(a_1,a_2,a_3)} + \frac{\partial(x_1,q_2,x_3)}{\partial(a_1,a_2,a_3)} + \frac{\partial(x_1,x_2,q_3)}{\partial(a_1,a_2,a_3)}.\end{aligned}$$

Writing each term out we see that

$$\frac{\partial(q_1,x_2,x_3)}{\partial(a_1,a_2,a_3)} = \begin{vmatrix} \frac{\partial q_1}{\partial a_1} & \frac{\partial q_1}{\partial a_2} & \frac{\partial q_1}{\partial a_3} \\ \frac{\partial x_2}{\partial a_1} & \frac{\partial x_2}{\partial a_2} & \frac{\partial x_2}{\partial a_3} \\ \frac{\partial x_3}{\partial a_1} & \frac{\partial x_3}{\partial a_2} & \frac{\partial x_3}{\partial a_3} \end{vmatrix}.$$

However

$$\frac{\partial q_1}{\partial a_i} = \frac{\partial q_1}{\partial x_1} \cdot \frac{\partial x_1}{\partial a_i} + \frac{\partial q_1}{\partial x_2} \cdot \frac{\partial x_2}{\partial a_i} + \frac{\partial q_1}{\partial x_3} \cdot \frac{\partial x_3}{\partial a_i},$$

and so using properties of the determinant we obtain

$$\frac{\partial(q_1,x_2,x_3)}{\partial(a_1,a_2,a_3)} = J \frac{\partial q_1}{\partial x_1}.$$

The other two terms can be treated similarly and hence

$$\frac{dJ}{dt} = J \operatorname{div} q. \tag{1.4}$$

We can now consider the rate of change of any property, such as total momentum, associated with a material volume $V(t)$, that is, a volume consisting of the same fluid particles at any time. By the continuum hypothesis

$$\begin{aligned} \frac{d}{dt}\left\{\int_{V(t)} L(x,t)dV(x)\right\} &= \frac{d}{dt}\left\{\int_{V(0)} L(x(a,t),t)JdV(a)\right\} \\ &= \int_{V(0)} \frac{d}{dt}(L(x(a,t),t)J)dV(a) \\ &= \int_{V(0)} (J\frac{dL}{dt} + LJ \operatorname{div} q)dV(a) \quad \text{(on using (1.4))} \\ &= \int_{V(t)} (\frac{dL}{dt} + L \operatorname{div} q)dV(x). \end{aligned} \tag{1.5}$$

This formula for "differentiating over a volume which is moving with the fluid" is called the *transport theorem*. We can now apply this theorem to derive the fundamental conservation laws.

The Equations of Inviscid Fluid Flow

Conservation of mass of any material volume $V(t)$ can be written

$$\frac{d}{dt}\int_{V(t)} \rho \, dV = 0$$

or, using (1.5),

$$\int_{V(t)} \left(\frac{d\rho}{dt} + \rho \operatorname{div} q\right) dV = 0.$$

Since V is arbitrary this leads to the differential equation

$$\frac{d\rho}{dt} + \rho \operatorname{div} q = 0, \tag{1.6}$$

which holds at each point of the fluid. In obtaining this equation it has been assumed that ρ and q are continuous and differentiable. We can deal with flows which contain discontinuities across a surface in x,t space, but equation (1.6) is not valid at the discontinuity. We shall return to this situation in Chapters III and V. For an incompressible fluid, ρ is constant and equation (1.6) reduces to

$$\operatorname{div} q = 0. \tag{1.7}$$

We next consider the linear momentum of the fluid contained in V(t). The forces acting on this volume are the internal surface forces acting on the boundary S of V together with any external body forces that may be acting. If we assume that the fluid is inviscid then the internal forces are just due to the pressure and act along the normal to the surface at every point. Suppose that there is an external body force F per unit mass and that we can apply Newton's equations to a fluid body. Then

$$\frac{d}{dt}\int_{V(t)} \rho q \, dV = -\int_{S(t)} p dS + \int_{V(t)} \rho F \, dV,$$

and using (1.5) on the left hand side and the divergence theorem on the surface integral we obtain

$$\int_{V(t)} \left(\frac{d}{dt}(\rho q) + \rho q \operatorname{div} q\right) dV = \int_{V(t)} (-\operatorname{grad} p + \rho F) dV.$$

Remembering that V(t) is arbitrary and using equation (1.6), we can reduce this to

$$\frac{dq}{dt} = F - 1/\rho \operatorname{grad} p, \tag{1.8}$$

which is Euler's equation for an inviscid fluid. If (1.6) and (1.8) are both satisfied it can be verified that the angular momentum of any volume V is also conserved.

For an incompressible fluid, equations (1.7) and (1.8) are sufficient to determine p and q, but for a compressible fluid we need another relation between p, ρ and T. This will be discussed in Section 4. In either case we need another equation involving T, which comes from considering conservation of energy.

The energy of an inviscid compressible fluid consists of two parts: the kinetic energy of the fluid particles and, in addition, the internal energy of the fluid. For an ideal gas, the internal energy per unit mass e is the total heat per unit mass, which is equal to C_vT†, where C_v is the specific heat at constant volume* and T is the temperature measured from absolute zero. The rate of change of energy in a material volume V must be balanced against

i) the rate at which work is done on the fluid by external forces and the pressure forces;

ii) the rate at which heat is gained across the boundary of V;

iii) the rate at which heat is created inside V by any source terms such as radiation.

The rate at which heat is conducted in a direction n is $(-k \text{ grad } T)\cdot n$ where k is the conductivity of the material.* Thus, conservation of energy for V leads to the equation

$$\frac{d}{dt}\left[\int_{V(t)} (\tfrac{1}{2}\rho q^2 + \rho e)dV\right]$$
$$= \int_{V(t)} \rho F\cdot q \, dV - \int_S pq\cdot dS + \int_S k \text{ grad } T\cdot dS + \frac{d}{dt}\left[\int_V \rho \, Q \, dV\right],$$

where Q is the heat addition per unit mass. Using the transport theorem and transforming the surface integrals by the divergence theorem, we obtain the equation

$$\rho q \cdot \frac{dq}{dt} + \rho \frac{de}{dt} = \rho F\cdot q - \text{div}(pq) + \text{div}(k \text{ grad } T) + \rho \frac{dQ}{dt},$$

and on using (1.8) and (1.6) we can reduce this to

$$\rho \frac{de}{dt} = p/\rho \frac{d\rho}{dt} + \text{div}(k \text{ grad } T) + \rho \frac{dQ}{dt}. \tag{1.9}$$

For incompressible flow, the energy equation becomes

$$\rho C_v \frac{dT}{dt} = k \nabla^2 T + \rho \frac{dQ}{dt}, \tag{1.10}$$

which is the *convective heat conduction equation* including a source term. If equations (1.7) and (1.8) can first be solved for q, then equation (1.10) is a linear equation for T.

---

†These results of thermodynamics will be discussed further in Section 4.

*C_v and k will be treated as constants throughout.

The boundary condition for an inviscid fluid at a fixed boundary has already been derived and is given by (1.2). A free boundary may occur between two immiscible fluids or between a fluid and a region of constant pressure and then, although (1.2) still holds, we need another equation to determine $f(x,t)$. This comes from applying the momentum principle to an element of fluid on the boundary, which shows that the pressure must be continuous across the boundary (as in Figure 1.1).

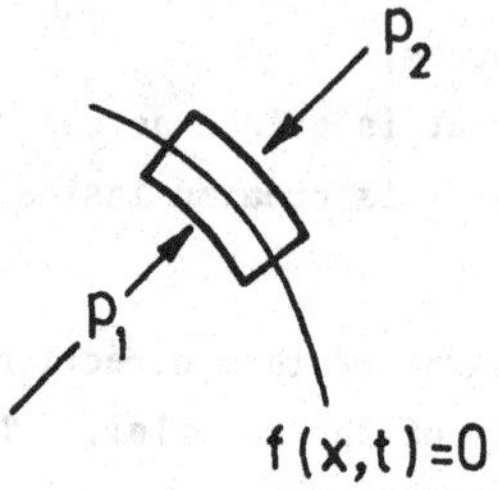

Figure 1.1. Conditions at a free surface.

To complete the model of an inviscid fluid two further observed phenomena should be noted. One is that the pressure is always positive. The other is that in reality the tangential velocity, as well as the normal velocity, is zero at a rigid boundary; we will discuss this restriction in the next section.

2. VORTICITY AND IRROTATIONALITY FOR INCOMPRESSIBLE FLOW

Vorticity

The *vorticity* ζ is defined by $\zeta = \text{curl } q$, so that $\text{div } \zeta = 0$. It is of interest to consider the evolution of ζ, and to do this we first write (1.8) in the form

$$\frac{\partial q}{\partial t} + \tfrac{1}{2} \text{ grad } q^2 - q \wedge \text{curl } q = F - 1/\rho \text{ grad } p.$$

If we assume that F is a conservative force (so that $F = -\text{grad } \Omega$) and take the curl of this equation, we obtain

$$\frac{d\zeta}{dt} = \frac{\partial \zeta}{\partial t} + (q\cdot\nabla)\zeta = (\zeta\cdot\nabla)q, \tag{1.11}$$

which is Helmholtz' equation. If we use Lagrangian variables, this equation can be integrated† to give Cauchy's equation

$$\zeta = (\zeta_0 \cdot \nabla_a) x(a,t), \tag{1.12}$$

where ∇_a is the gradient operator with respect to Lagrangian variables a, and ζ_0 is the value of ζ at $t = 0$ when $x = a$.

This equation represents the physical fact that vorticity is constant over a stream tube which is moving with the fluid. From this we see that if the vorticity is everywhere zero in a fluid region $V(0)$ at time $t = 0$, then it will be zero for all subsequent times in the region $V(t)$ which is occupied at that time by the fluid which was in $V(0)$ at $t = 0$. Then $\text{curl } q \equiv 0$ in $V(t)$ and we have *irrotational* flow. Such flows occur, for example, when the fluid is initially at rest or when there are uniform conditions at infinity.

<u>Kelvin's Theorem</u>

An alternative way of thinking about vorticity is to consider the total vorticity flux through an arbitrary closed contour C in a simply connected region of the fluid. This is called the *circulation* round C and is given by

---

†Using suffix notation and the summation convention, (1.11) is written as

$$\frac{d\zeta_i}{dt} = \zeta_j \frac{\partial q_i}{\partial x_j} = \zeta_j \frac{\partial q_i}{\partial a_k}\frac{\partial a_k}{\partial x_j} = \zeta_j \frac{\partial a_k}{\partial x_j}\frac{d}{dt}\left(\frac{\partial x_i}{\partial a_k}\right).$$

Hence

$$\frac{d}{dt}\left(\zeta_i \frac{\partial a_\ell}{\partial x_i}\right) = \zeta_j \frac{\partial a_k}{\partial x_j}\frac{\partial a_\ell}{\partial x_i}\frac{d}{dt}\left(\frac{\partial x_i}{\partial a_k}\right) + \zeta_i \frac{d}{dt}\left(\frac{\partial a_\ell}{\partial x_i}\right)$$

$$= \zeta_j \frac{\partial a_k}{\partial x_j}\frac{d}{dt}\left[\frac{\partial a_\ell}{\partial x_i}\frac{\partial x_i}{\partial a_k}\right], \quad \text{on using} \quad \frac{\partial a_k}{\partial x_j}\frac{\partial x_i}{\partial a_k} = \delta_{ij},$$

$$= 0 \qquad\qquad, \text{ on using } \quad \frac{\partial a_\ell}{\partial x_i}\frac{\partial x_i}{\partial a_k} = \delta_{\ell k}.$$

Thus $\zeta_i \frac{\partial a_\ell}{\partial x_i}$ is constant for a fluid particle and $\zeta_i \frac{\partial a_\ell}{\partial x_i} = \zeta_{0\ell}$, or $\zeta_j = \zeta_{0\ell}\frac{\partial x_j}{\partial a_\ell}$, which is (1.12).

$$\Gamma = \int_C q\cdot dx = \iint_S \zeta\cdot dS \tag{1.13}$$

by Stokes' Theorem. The circulation is defined by the line integral $\int q\cdot dx$ in a region which need not be simply connected and where the surface integral may not exist. If we want to consider changes in Γ when $C(t)$ is a closed curve moving with the fluid, we need again to resort to Lagrangian variables. From (1.13),

$$\Gamma = \int_{C(t)} q_i dx_i = \int_{C(0)} q_i \frac{\partial x_i}{\partial a_j} da_j,$$

and

$$\frac{d\Gamma}{dt} = \int_{C(0)} \frac{d}{dt}\left(q_i \frac{\partial x_i}{\partial a_j}\right) da_j$$
$$= \int_{C(t)} \frac{dq}{dt}\cdot dx + \int_{C(t)} q\cdot dq.$$

Now $\int_C q\cdot dq = [\frac{1}{2} q^2]_C = 0$ since q is a single valued function of x and t, and using (1.8) we have

$$\frac{d\Gamma}{dt} = \int_C F\cdot dx \tag{1.14}$$

since p is also a single valued function of x and t. If the body forces are conservative with $F = -\text{grad}\,\Omega$, then $\frac{d\Gamma}{dt} = 0$. This gives *Kelvin's Theorem*, that the circulation around any contour which moves with the fluid is constant. In particular, we again arrive at the result that if $\zeta \equiv 0$ at $t = 0$ then $\Gamma = 0$ for all closed curves and so $\zeta \equiv 0$ for all time.

The Velocity Potential

For irrotational flow we can define a *velocity potential* ϕ by $\int_\infty^x q\cdot dx$. From Kelvin's theorem ϕ will be a single valued, continuous function of x and t.

Taking the gradient of this integral leads to

$$q = \text{grad}\,\phi, \tag{1.15}$$

and the equations of flow for an incompressible fluid, (1.7) and (1.8), become

$$\nabla^2\phi = 0 \tag{1.16}$$

and

$$\frac{\partial\phi}{\partial t} + \frac{1}{2} q^2 + \Omega + p/\rho = G(t), \tag{1.17}$$

where G is an arbitrary function of t. Equation (1.17) is *Bernoulli's equation*. The problem is therefore reduced to two scalar equations; first we solve Laplace's equation (1.16) for ϕ and then we find p from (1.17) after determining G from the boundary conditions. Because Laplace's equation is elliptic, only one condition on ϕ can be imposed on the boundaries. For a solid boundary we must prescribe the normal velocity $\frac{\partial \phi}{\partial n}$, but in general there will be a slip velocity q_t tangential to the boundary, which is contrary to the required physical condition. Thus, the model is not adequate to describe flows with rigid boundaries without some modification.

Boundary Layers

Briefly, we observe that for many flow situations the tangential velocity changes very rapidly in the direction normal to the wall, from zero at the wall to the slip velocity q_t. In this *boundary layer* viscous forces cannot be neglected; their action will be considered in Section 5 of this chapter. However in many situations these layers are very thin and can be modelled by a discontinuity in the tangential velocity. Such a discontinuity is called a *shear layer* and contains concentrated vorticity. For example, in two-dimensional flow past a flat plate the discontinuity in velocity can be represented by a line of concentrated vorticity of strength q_t on the plate, as shown in Figure 1.2. For two-dimensional

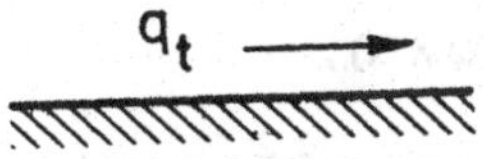

Figure 1.2. Conditions at a fixed surface.

flow (1.11) reduces to $\frac{d\zeta}{dt} = 0$, and vorticity will be convected with the fluid. Thus for flow past a finite body, vorticity will be continuously created on the boundary and convected downstream to form a wake. The irrotational flow model is only relevant for flows with free boundaries or flows with fixed boundaries which are "streamlined" so that the shear layer remains thin and any wake which forms is also thin. For bodies that are not streamlined, the flow will still be irrotational outside the boundary layer and the wake, but if the wake is thick it is difficult to construct a realistic rotational model of flow within the wake.

Solutions of Laplace's equation in irrotational incompressible flow situations are described in many books on hydromechanics. We shall only be concerned with such solutions for problems with free boundaries, in Chapters 2, 3 and 7, and with one particular application of complex variable theory in Chapter 7. Meanwhile, to demonstrate the special nature of irrotational flows we shall examine models for flows which are rotational.

3. ROTATIONAL INCOMPRESSIBLE FLOWS

Rotating Flows

A simple example of a rotational flow occurs when the frame of reference of the fluid is rotating. In meteorological or oceanographic problems the effects of earth's rotation may have to be included; these are examples of motions relative to a rotating frame. On a laboratory scale, if a container of fluid is rotated for a long time the viscous forces will eventually bring about rigid body rotation of the fluid. Then, relative to the container, any disturbance will produce a rotational flow in a fluid initially at rest. Consider a system rotating with constant angular velocity ω, let $\frac{D}{Dt}$ denote differentiation following the fluid with respect to fixed axes, and let $\frac{d}{dt}$ and $\frac{\partial}{\partial t}$ have their usual meaning with respect to the rotating system. Then the fluid acceleration is

$$\frac{D^2 x}{Dt^2} = \left(\frac{d}{dt} + \omega \wedge\right)\left(\frac{dx}{dt} + \omega \wedge x\right)$$

$$= \frac{dq}{dt} + 2\omega \wedge q + \omega \wedge (\omega \wedge x),$$

where q is the velocity relative to the rotating frame. Relative to the rotating frame, Euler's equations (1.8) become

$$\frac{dq}{dt} + 2\omega \wedge q = -1/\rho \operatorname{grad} \bar{p}, \tag{1.18}$$

where

$$\bar{p} = p + \rho\Omega - \tfrac{1}{2}\rho(\omega \wedge x)^2 \tag{1.19}$$

is the reduced pressure. The continuity equation for an incompressible fluid (1.7) is unchanged. Instead of Kelvin's theorem, we have

$$\frac{d\Gamma}{dt} = -2\omega \cdot \int_C q \wedge dx, \tag{1.20}$$

so that in general Γ varies with time. The vorticity now satisfies the equation

$$\frac{d\zeta}{dt} = (\zeta\cdot\nabla)q + 2(\omega\cdot\nabla)q, \tag{1.21}$$

and ζ will be non-zero even for flows starting from rest relative to the rotating frame. For a two dimensional motion, however, in which $\omega = (0,0,\omega)$ and $q = (q_1(x_1,x_2),q_2(x_1,x_2),0)$, $\zeta = (0,0,\zeta(x_1,x_2))$ and, since the right hand side of equation (1.21) is zero, ζ will be constant for a fluid particle.

For a detailed discussion of solutions of these equations reference should be made to Greenspan [13].

Shear Flows

Vorticity may be introduced into the mathematical model of a flow through the boundary conditions. The simplest example is a flow due to a nonuniform stream such as a shear flow at infinity in which the velocity distribution varies linearly across the stream. For simplicity we consider two dimensional problems in which the vorticity has one component: $\zeta = (0,0,\zeta)$. This problem may be simplified by the introduction of the two dimensional incompressible *stream function* ψ defined by

$$\psi(x,t) = \int_0^x q_n ds, \tag{1.22}$$

where 0 is some reference point, the integral is taken along any path C from 0 to x, and q_n is the component of q normal to C. Thus ψ represents the volume flow between 0 and x and hence is a uniquely defined function independent of the path C. From definition (1.22)

$$q = (\frac{\partial\psi}{\partial x_2}, -\frac{\partial\psi}{\partial x_1}, 0) \tag{1.23}$$

and the continuity equation (1.7) is identically satisfied.† An instantaneous streamline is defined as a curve such that the velocity at each point on it is tangential to the curve at a given time t. Thus there is no flow across a streamline, and in two dimensional flow there exists a family of streamlines given by lines on which ψ is constant. In an unsteady flow the streamlines change with time and are different from the particle paths defined on page 1.

From the definition of vorticity and (1.23)

†A more general form of this result is that since $\operatorname{div} q = 0$ we can find a vector potential A such that $q = \operatorname{curl} A$ and $\operatorname{div} A = 0$. In this two dimensional case, $A = (0,0,\psi)$.

$$\zeta = -\left(\frac{\partial^2\psi}{\partial x_1^2} + \frac{\partial^2\psi}{\partial x_2^2}\right). \tag{1.24}$$

From (1.11), $\frac{d\zeta}{dt} = 0$ in two dimensional flow, which may be written

$$\frac{\partial\zeta}{\partial t} = \frac{\partial(\psi,\zeta)}{\partial(x_1,x_2)}. \tag{1.25}$$

For steady flow, $\frac{\partial(\psi,\zeta)}{\partial(x_1,x_2)} = 0$ and ψ,ζ are functionally related by $\zeta = F(\psi)$, where F is some differentiable function. Thus, a class of steady rotational two-dimensional flows are given by solutions of (1.24) with various forms of $F(\psi)$. However, an explicit solution is only practicable when $F(\psi)$ is a constant or a linear function of ψ. The simplest example is a constant shear flow at infinity so that at infinity $q = (\kappa x_2,0,0)$ and $\zeta = -\kappa$ is constant. Then $\zeta = -\kappa$ everywhere, and (1.24) can be solved by the usual techniques available for Poisson's equation. A second example is two dimensional flow in which the fluid is rotating with constant angular velocity at infinity. This has already been discussed in the previous section.

To calculate the pressure for a steady rotational flow we can rewrite (1.8) to give

$$\text{grad}(p/\rho + \tfrac{1}{2}q^2 + \Omega) = q \wedge \zeta. \tag{1.26}$$

The right hand side is perpendicular to the streamline at every point and $(p/\rho + \frac{1}{2}q^2 + \Omega)$ is constant along a streamline. In the two dimensional case

$$q \wedge \zeta = -\zeta \text{ grad } \psi = -F(\psi) \text{ grad } \psi.$$

Then, integrating (1.26) leads to

$$p/\rho + \tfrac{1}{2}q^2 + \Omega + \int_0^\psi F(\psi')d\psi' = \text{constant}. \tag{1.27}$$

This is a modified form of Bernoulli's equation.

Flows of an Inhomogeneous Fluid

Another example of a rotational flow occurs for a non-homogeneous fluid. An incompressible non-homogeneous fluid is one for which $\frac{d\rho}{dt} = 0$ but at $t = 0$ the density varies in space so that $\rho(x,0) = \rho(a)$ is not constant. Hence

$$\frac{\partial\rho}{\partial t} + q \cdot \text{grad } \rho = 0, \tag{1.28}$$

but in addition the continuity equation (1.6) requires that $\operatorname{div} q = 0$. For any flow with variable density we can take the curl of equation (1.8) to obtain

$$\frac{d\zeta}{dt} - (\zeta\cdot\nabla)q + \zeta\cdot\operatorname{div} q = 1/\rho^2 \operatorname{grad}\rho \wedge \operatorname{grad} p,$$

and using (1.6) we can rewrite this as

$$\frac{d}{dt}\left(\frac{\zeta}{\rho}\right) = 1/\rho\,(\zeta\cdot\nabla)q + 1/\rho^3 \operatorname{grad}\rho \wedge \operatorname{grad} p. \tag{1.29}$$

For steady flows we can again integrate (1.8) along a streamline, using (1.24), to obtain

$$\tfrac{1}{2}q^2 + p/\rho + \Omega = \text{constant on a streamline.}$$

From (1.29), we see that vorticity is produced if $\operatorname{grad}\rho \wedge \operatorname{grad} p \neq 0$. If $\operatorname{grad}\rho \wedge \operatorname{grad} p$ is zero then p is a function of ρ and t only, and the flow is said to be *barotropic*. In general the flow of a stratified fluid is not barotropic.

An example of a non-homogeneous situation is that of a fluid stratified due to the effects of gravity. In a realistic model for disturbances in the atmosphere or the ocean, the configuration at time $t = 0$ will be a vertical density variation so that $\frac{\partial\rho}{\partial x_3} \neq 0$. For a steady horizontal flow the solution $\rho = \rho(x_3)$, $q = (q_1, q_2, 0)$ is possible, and the flow consists of layers of irrotational two dimensional flows with different densities. We shall not discuss this situation further; for more details reference should be made to Yih [27].

4. COMPRESSIBLE FLOW AND THERMODYNAMICS

When we consider a compressible fluid we find that equations (1.6), (1.8) and (1.9) are insufficient to determine q, ρ, p and T, and another relation between the variables of state is needed. If the fluid is barotropic there is a pressure-density law of the form $p = p(\rho,t)$. Equation (1.29) then reduces to equation (1.11) with ζ replaced by ζ/ρ, and we can again deduce that all flows which are initially irrotational will remain irrotational. Hence a velocity potential exists for a compressible fluid or gas, and equation (1.8) integrates to give the compressible form of Bernoulli's equation:

$$\frac{\partial\phi}{\partial t} + \tfrac{1}{2}q^2 + \int\frac{dp}{\rho} + \Omega = G(t), \tag{1.30}$$

where $\int \frac{dp}{\rho}$ means $\int \frac{\partial p}{\partial \rho} \frac{d\rho}{\rho}$. Kelvin's Theorem remains unaltered. To discover whether a compressible fluid is barotropic and which pressure-density law is appropriate, we must make use of results from Thermodynamics. This is a complicated subject for a general gas which is not in thermodynamic equilibrium, and we shall only discuss what we need for our purpose. For a fuller account reference should be made to Liepmann and Roshko [17]. We shall only consider gases in thermodynamic equilibrium, which implies that the time scale of the macroscopic changes in which we are interested is much longer than the thermodynamic time scales. The "thermodynamic state" of a gas in equilibrium can be defined by any two of the variables e, p, ρ and T, so that both e and p are functions of ρ and T. For a *perfect gas* we postulate the gas law

$$p = R\rho T, \tag{1.31}$$

where R is the gas constant and T is the temperature measured from absolute zero. This law can be verified experimentally for a gas at rest and we assume that it will also apply to a gas in motion. In addition we assume that

$$e = C_v T, \tag{1.32}$$

where C_v is constant. This may be verified experimentally and is found to be valid for most perfect gases which are not at extreme temperatures.

A perfect gas in which heat conduction may be neglected is called an *ideal gas*. With $k = 0$ in equation (1.9) we obtain

$$\frac{de}{dt} + p \frac{d}{dt} \left[\frac{1}{\rho}\right] = \frac{dQ}{dt}.$$

We can now define the *entropy* S by

$$T \frac{dS}{dt} = \frac{dQ}{dt} = \frac{de}{dt} + p \frac{d}{dt} (1/\rho). \tag{1.33}$$

The statement $T \frac{dS}{dt} = \frac{dQ}{dt}$ is a macroscopic version of the Second Law of Thermodynamics that $T\,dS = \Delta Q$, which ensures that a scalar function S exists just as the first law of Thermodynamics ensures that a scalar function e exists. Using (1.31) and (1.32), we see that equation (1.33) becomes

$$\frac{dS}{dt} = \frac{C_v}{T} \frac{dT}{dt} + R\rho \frac{d}{dt} (1/\rho),$$

and

$$S = C_v \log T + R \log (1/\rho) + \text{constant}. \tag{1.34}$$

When there is no internal heat source term, $Q = 0$, and the energy equation (1.33) reduces to

$$\frac{dS}{dt} = 0. \tag{1.35}$$

Thus for an ideal gas, the entropy of each fluid particle is constant and the flow is *isentropic*. This may be contrasted with an incompressible non-conducting fluid where the temperature is conserved for a fluid particle from (1.10).

If the entropy of all the fluid particles is the same constant for all time then the flow is *homentropic*. It is now easy to show that homentropic flows of an ideal gas are barotropic. If we write $R = (\gamma-1)C_v$, where γ is a constant called the ratio of the specific heats, then from (1.34) and (1.31)

$$p = \rho^{\gamma} \exp\left(\frac{S - S_0}{C_v}\right), \tag{1.36}$$

where S_0 is some suitable constant. In a homentropic flow S is constant everywhere, and the appropriate barotropic law is that p is proportional to ρ^{γ}. Bernoulli's equation (1.30) then becomes

$$\frac{\partial\phi}{\partial t} + \tfrac{1}{2}q^2 + \frac{\gamma p}{(\gamma-1)\rho} + \Omega = G(t) \tag{1.37}$$

for a homentropic gas. If the flow is isentropic, but not homentropic, we can still obtain the Bernoulli integral for a steady flow since S is constant along a streamline. In this case, equation (1.8) becomes

$$\operatorname{grad}\left(\tfrac{1}{2}q^2 + \Omega + \int \frac{dp}{\rho}\right) = q \wedge \zeta \tag{1.38}$$

and

$$\tfrac{1}{2}q^2 + \Omega + \int \frac{dp}{\rho} = F(S). \tag{1.39}$$

The function $\int \frac{dp}{\rho}$ is called the *enthalpy*. The integration is carried out keeping S constant, where p is a function of S and ρ as in (1.36). For non-isentropic flows of an ideal gas, the enthalpy is defined as $h = e + p/\rho = C_pT$, where C_p is the specific heat at constant pressure and $C_p = \gamma C_v$. From (1.38) with $h = h(S,\rho)$ we have, after some manipulation,

$$q \wedge \zeta = \left(\frac{\partial F}{\partial S} - \frac{\partial h}{\partial S}\right)\operatorname{grad} S. \tag{1.40}$$

This is *Crocco's Theorem*. It shows how vorticity is created in steady non-homentropic flows. It is an integral of equation (1.29), the function $F(S)$ being determined from the boundary conditions.

5. VISCOUS FLOW

As we have already seen in Section 2, there are often regions of the flow field where viscous effects are important. Even though these regions may be very thin, it is necessary to understand their structure in order to be sure we are applying the correct boundary conditions to the inviscid model. We now formulate briefly the model equations for an incompressible viscous flow. For a more detailed account of this subject reference may be made to Batchelor [1].

When viscosity is taken into consideration, the internal forces are represented by a stress tensor τ_{ij}. This is defined so that the force per unit area on an element of surface with unit normal (n_1,n_2,n_3) is $n_i\tau_{ij}e_j$, where summation over i and $j = 1,2,3$ is implied. Thus τ_{ij} is the force per unit area on a plane perpendicular to the x_i direction and acting in the x_j direction. In cartesian coordinates τ_{ij} has the form $-p\delta_{ij} + \sigma_{ij}$, where σ_{ij} is a symmetric tensor. A *Newtonian fluid* is one for which there is a linear dependence between σ_{ij} and the rate of strain tensor $e_{ij} = \frac{1}{2}(\partial q_i/\partial x_j + \partial q_j/\partial x_i)$. Arguments of symmetry and the invariance of the stress tensor under linear translations and rigid body rotations then lead to the stress-strain relation in the form

$$\sigma_{ij} = 2\mu e_{ij} + \lambda\delta_{ij} \operatorname{div} q, \tag{1.41}$$

where λ,μ are coefficients of viscosity and are here assumed to be constant.†

The conservation of momentum equation now becomes

$$\rho \frac{dq_i}{dt} = \rho F_i - \frac{\partial p}{\partial x_i} + \frac{\partial \sigma_{ij}}{\partial x_j}, \tag{1.42}$$

and in the incompressible case when $\operatorname{div} q = 0$ and μ is constant, this reduces to the *Navier-Stokes* equation

$$\frac{dq}{dt} = F - 1/\rho \operatorname{grad} p + \nu\nabla^2 q, \tag{1.43}$$

---

†In practice λ and μ may be sensitive to temperature changes but we assume that they do not change significantly in the temperature range in which we are interested.

where $\nu = \mu/\rho$ is the kinematic viscosity. If we now take $\nu = 0$ in (1.43) we obtain the inviscid Euler equation (1.8), but the term containing the highest derivatives had disappeared. We cannot therefore expect to satisfy as many boundary conditions with $\nu = 0$ as with $\nu \neq 0$, and a non-zero tangential velocity at a rigid boundary is almost inevitable in an inviscid flow. When ν is very small the viscous term $\nu\nabla^2 q$ will only be comparable with the inertia terms $\frac{dq}{dt}$ if the second derivative of q is large. This crude argument, which is capable of much refinement, leads to the concept of a boundary layer whose thickness is proportional to $\nu^{\frac{1}{2}}$, across which there is a finite change in the velocity tangential to the layer. In the limit as $\nu \to 0$ this layer becomes the shear layer which has already been mentioned. As long as the boundary layers and any resulting wakes in a problem remain thin, the inviscid model will be relevant. We shall not discuss boundary layer theory here. Reference may be made to Rosenhead [21] for a detailed discussion. The energy equation for a viscous fluid is

$$\rho T \frac{dS}{dt} = k\nabla^2 T + \sigma_{ij} \frac{\partial q_i}{\partial x_j}, \tag{1.44}$$

where the final term is summed over i and $j = 1,2,3$. This term, called the *dissipation*, represents the work done by the viscous forces and is never negative.

One further postulate from Thermodynamics is needed, namely, that entropy cannot decrease; that is,

$$\frac{d}{dt}\left\{\int_{V(t)} S \, dV\right\} \geq 0. \tag{1.45}$$

However, for an inviscid fluid with negligible heat conduction, S is constant following the fluid and statement (1.45) is superfluous.

6. DIMENSIONAL ANALYSIS

We are concerned with deriving mathematical models of certain physical phenomena, and in any practical problem it is necessary to decide which physical features are important and must be included. This can be done intuitively, but more quantitative criteria are useful and can be obtained by dimensional methods. For a given problem we can usually identify a suitable time and length scale or, what is equivalent, a reference length L and a reference velocity U. We can then define non-dimensional variables

$$\hat{x} = \frac{x}{L}, \quad \hat{q} = \frac{q}{U}, \quad \hat{t} = \frac{tU}{L},$$

and attempt to estimate the relative importance of any two physical features of the flow. As an example, consider the flow of an inviscid incompressible fluid with an external body force due to gravity. The non-dimensional forms of equations (1.7) and (1.8) are then

$$\nabla \cdot \hat{q} = 0; \qquad \frac{d\hat{q}}{d\hat{t}} = -\nabla\hat{p} + \frac{gL}{U^2} k,$$

where $\hat{p} = p/\rho U^2$, k is a unit vector in the vertical direction, and ∇ is the gradient operator in $\hat{x}$ coordinates. There is a non-dimensional parameter gL/U^2 in these equations, and a range of problems will have "similar" solutions if they correspond to the same value of this parameter. The effects of gravity will be small compared to those of inertia if this parameter is small, that is, if the *Froude number* $F = \dfrac{U}{(gL)^{\frac{1}{2}}}$ is very large. If we omit gravity effects altogether, as we often do, then we are finding the first term in an asymptotic expansion of the solution for large F, under the assumption that such an expansion exists as a function of F as $F \to \infty$.

For a rotating flow governed by equation (1.18) we can estimate the relative importance of the convective and rotational effects. With $\omega = \omega k$ we could use either L/U or $1/\omega$ as the time scale. If we are interested in a time scale of the order of a period of rotation as, for example, in a meteorological problem, we naturally define $\hat{t} = \omega t$. Then the equations become

$$\nabla \cdot \hat{q} = 0, \qquad \frac{\partial\hat{q}}{\partial\hat{t}} + \varepsilon(\hat{q}\cdot\nabla)\hat{q} + 2k \wedge \hat{q} = -\frac{\nabla p}{\rho L\omega U}, \tag{1.46}$$

where $\varepsilon = U/\omega L$ is called the *Rossby Number*. Convective effects will be unimportant if $\varepsilon \ll 1$, and in this case the correct scaling for the pressure is $\rho L\omega U$. It is likely that ε is small in meteorological problems where the length scale L is usually large; in the laboratory it requires rapid rotation to make ε small. The scaling for the pressure often cannot be determined from the boundary conditions and must then be chosen so that the pressure term does not disappear in the first order approximate equations, since this would lead to more equations than unknowns. When $\varepsilon \gg 1$, the appropriate time scale for this problem is L/U, and the scaling for p will now be $\varepsilon\rho L\omega U = \rho U^2$. Then we can see that the convective terms balance the pressure gradient and the effects of rotation may be ignored.

Asymptotic expansions of this kind in a small (or large) parameter are not always uniformly valid even if they exist; that is, the asymptotic solution obtained by putting the parameter equal to zero (or infinity) may not be valid for all x and t. The classic example is to compare the effects of inertia and viscosity for an incompressible fluid. In non-dimensional form the Navier-Stokes equation (1.43) becomes

$$\frac{d\hat{q}}{d\hat{t}} = -\nabla p + \frac{1}{Re}\nabla^2\hat{q}, \tag{1.47}$$

where $Re = UL/\nu$ is the *Reynolds number*. Viscous effects should be unimportant if $Re \gg 1$, but the asymptotic expansion for large values of Re is not uniform and the solution with $1/Re = 0$ is only valid away from rigid boundaries. The problem is said to be *singular* near the rigid boundary, where there is a boundary layer. There are technical difficulties involved in dealing with singular perturbations of this type and the associated ideas of matched asymptotic expansions. We shall not pursue these ideas any further, but they are discussed in detail in Van Dyke [25].

Other parameters that often occur in flow problems are the Mach number associated with compressibility, the Prandtl number with heat conduction, the Nusselt number with surface tension and the Ekman number associating the effects of rotation and viscosity. In any given problem the first step is to decide on the correct scales for the dependent and independent variables and to evaluate the nondimensional parameters that arise in connection with the various possible physical effects. If the parameter associated with a particular effect is small and if the solution can be expanded asymptotically in that parameter, then it is appropriate to omit that physical effect from the model to obtain a first approximation. In regions in which the asymptotic expansion is not valid this approximation will not be valid. The models developed in this chapter are all valid only in situations in which the omitted physical features satisfy the above criteria.

EXERCISES

1. (i) From the transport theorem (1.5) show that the rate of change of the kinetic energy of an inviscid compressible fluid contained in a volume V(t) is

$$-\int_S p\mathbf{q}\cdot d\mathbf{S} + \int_V p \operatorname{div} \mathbf{q}\, dV,$$

where S is the boundary of V.

(ii) Generalise the theorem to include possible discontinuities in L across an open surface Σ contained in $V(t)$ by the addition of the term $\int_\Sigma [L]_1^2 q\cdot n\, dS$, where $[L]_1^2$ is the discontinuity in L in crossing Σ in the direction of the normal n.

(iii) By choosing V at time t to consist of the fluid particles in a thin layer containing Σ, show that

$$[\rho u]_1^2 = 0 = [v]_1^2,$$

where u and v are velocity components normal and tangential to Σ. Obtain an expression for $[p]_1^2$.

2. Show that the rate of change of the angular momentum of an arbitrary volume of inviscid fluid is exactly equal to the couple applied by the pressure and body forces if the continuity and linear momentum equations are satisfied. If the fluid is viscous show that the principle of conservation of angular momentum implies that the stress tensor is symmetric.

3. (i) Show that the kinetic energy of an inviscid incompressible flow for which q_n is prescribed over the fluid boundary, together with any sources or sinks, is a minimum when the flow is irrotational.

(ii) A sphere of radius a is moving with speed U in fluid which is at rest far from the sphere. Show that the kinetic energy of the fluid is $\frac{1}{2}M'U^2$, where $M' = (2/3)\pi\rho a^3$ is called the *virtual mass* of the sphere.

4. (i) A vortex line is defined to be a line whose tangent at every point is parallel to the local vorticity. Using (1.12) or otherwise, show that in an incompressible fluid vortex lines move with the fluid.

(ii) Show that (1.12) may be generalized for a barotropic fluid to $\zeta = (\rho/\rho_0)(\zeta_0\cdot\nabla_a)x(a,t)$. Does 4(i) remain true in this case?

(iii) At time $t = 0$, a vortex of strength Γ is at a distance c from the centre of a fixed circular cylinder of radius a $(< c)$ in an otherwise unbounded incompressible fluid. Show that at time $4\pi^2c^2(c^2-a^2)/\Gamma a^2$ the flow pattern is unchanged.

5. A uniform incompressible stream of speed U flows past a circular cylinder, radius a, which is slightly porous. If the normal velocity at the surface $q_n = c(p-p_c)$, where p_c is constant, and the total flow into the cylinder is zero, look for a solution in the form $\phi = \phi_0 + c\phi_1 + O(c^2)$. Obtain an expression for ϕ_0 and show that $p_c = p_\infty - \frac{1}{2}\rho U^2$, where p_∞ is the pressure in the stream. Show also that

$$\phi_1 = -\rho \frac{U^2 a^3}{r^2} \cos 2\theta, \quad \text{(where } r \text{ and } \theta \text{ are polar coordinates)}$$

and sketch a streamline pattern.

6. A two-dimensional steady incompressible flow has velocity $q = (\alpha y, 0)$ as $x \to -\infty$. Show that the stream function ψ satisfies the equation

$$\frac{\partial^2 \psi}{\partial x^2} + \frac{\partial^2 \psi}{\partial y^2} = \alpha,$$

and obtain an expression for ψ for the flow of this stream past the circular cylinder $x^2 + y^2 = a^2$ in the form

$$\frac{\alpha}{4}\left[r^2 - a^2 - (r^2 - \frac{a^4}{r^2}) \cos 2\theta\right],$$

where r and θ are polar coordinates. Show that the pressure on $\theta = 0$ is given by

$$p - p_\infty = -\rho \frac{\alpha^2}{8}\left(2r^2 + a^2 + \frac{5a^4}{r^2} + \frac{a^8}{r^6}\right).$$

What is the equation satisfied by ψ if $q = (\alpha y^2, 0)$ as $x \to -\infty$?

7. Prove Kelvin's Theorem for a barotropic fluid in the form due to Bjerknes:

$$\frac{d\Gamma}{dt} = \int_C T\,dS, \text{ where } T \text{ is the temperature and } S \text{ the entropy.}$$

Show also that the helicity H, defined by

$$H = \int q \cdot \zeta \, dV,$$

the integral being taken over all the fluid, is constant in time if $\zeta \cdot n = 0$ on the boundary.

8. For a rotating incompressible fluid the Rossby number is defined by $U/\omega L$. If the Rossby number is small and may be neglected, show that for a steady flow $\partial q/\partial x_3 = 0$, where x_3 is measured in the direction of ω. Rotating fluid is contained in $0 \le x_3 \le h$, and the lateral velocity is made zero at a point on $x_3 = 0$ (by a small obstruction). Show that the velocity will be zero at this point for all x_3 and that the fluid flows two-dimensionally round this *Taylor column*, which is stationary with respect to the container.

9. For an unsteady flow with zero Rossby number, show that a solution may be found in the form $q = \mathcal{Q}(x)\rho^{i\lambda t}$, $p = \phi(x)\rho^{i\lambda t}$. For flows in a two dimensional container as in question 8, show that ϕ satisfies

$$\nabla^2\phi = \frac{4}{\lambda^2}\frac{\partial^2\phi}{\partial x_3^2}\,.$$

For a circular container $x_1^2 + x_2^2 = r^2 = a^2$, show that a possible motion is given by

$$\phi = J_0\left[(4-\lambda^2)^{\frac{1}{2}}\frac{n\pi r}{\lambda h}\right]\cos\frac{n\pi x}{h}\,,$$

provided $\lambda < 2$ and satisfies $J_0'\left[(4-\lambda^2)^{\frac{1}{2}}\frac{n\pi a}{\lambda h}\right] = 0$, where n is an integer. Motions satisfying these equations are called *inertial waves.*

10. A stratified fluid is at rest with $\rho = \rho_0(x_3)$. If the steady state is given a small perturbation so that the density is $\rho_0(x_3) + \rho(x,t)$ and the pressure is $p_0(x_3) + p(x,t)$, where ρ and p are small compared to ρ_0 and p_0, show that it is possible to find a solution for p in the form

$$p = f(x_3)\rho^{i(\lambda t+\ell x_1+mx_2)}\,,$$

where f satisfies

$$\rho_0 c\, f'' + \rho_0 c' f' - (\ell^2+m^2)f = 0$$

and

$$c = \frac{\lambda^2}{\lambda^2\rho_0 + g\rho_0'}\,.$$

Motions satisfying these equations are called *internal waves.*

Chapter II
Free Boundary Problems

1. FLOWS WITHOUT GRAVITY

One example of a flow with a free boundary is that of a jet of fluid travelling through a region of constant pressure. There are two typical situations which are shown in Figure 2.1. The first is a jet impinging on a fixed wall and the second is a jet emerging from a hole in the wall of a large reservoir. These situations may either be two or three-dimensional, but we can make more analytical progress in the two-dimensional case.

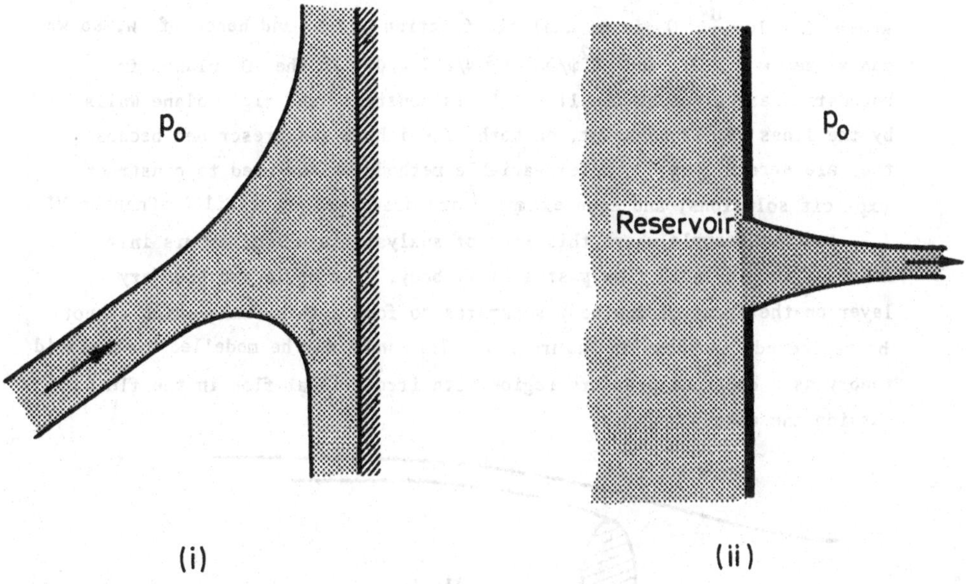

Figure 2.1. Jet flows.

If the fluid is inviscid and incompressible and there are no body forces, then from Bernoulli's equation (1.17) for steady flow the velocity, q, will be constant on the boundary of the jet. Also on the fixed rigid boundaries of the flow we know the direction of the velocity. It is therefore convenient to use the components of $\mathbf{q}$ as the independent variables since the position of both the free and the rigid boundaries is known in these variables. For two-dimensional irrotational flow we can find a suitable transformation; it is called a *hodograph transformation*, and the (q_1, q_2) plane is called the hodograph plane.

Since $q_1 = \frac{\partial\phi}{\partial x_1} = \frac{\partial\psi}{\partial x_2}$ and $q_2 = \frac{\partial\phi}{\partial x_2} = -\frac{\partial\psi}{\partial x_1}$, we may define a complex potential $w = \phi + i\psi$ which is an analytic function of $z = x_1 + ix_2$. If the velocity vector $\mathbf{q}$ has magnitude q and makes an angle θ with the x_1 axis, then

$$\frac{dw}{dz} = q_1 - iq_2 = qe^{-i\theta}, \tag{2.1}$$

and we define

$$Q = \log(U\,\frac{dz}{dw}) = L + i\theta, \tag{2.2}$$

where $L = \log\frac{U}{q}$. Q is an analytic function of z and hence of w, so we can write $w = w(Q)$ and $\partial^2\psi/\partial L^2 + \partial^2\psi/\partial\theta^2 = 0$. In the Q plane, free boundaries are given by the lines L is constant and rigid plane walls by the lines θ is constant, on both of which ψ is prescribed because they are streamlines. Complex variable methods may be used to construct explicit solutions, and some examples are discussed in detail in Chapter VII.

Another example where this kind of analysis is useful occurs in a model of the inviscid flow past a bluff body. The neglected boundary layer on the front of the body separates to form a thick wake which cannot be neglected, as shown in Figure 2.2. This wake may be modelled by inviscid theory as a constant pressure region with irrotational flow in the fluid outside the wake.

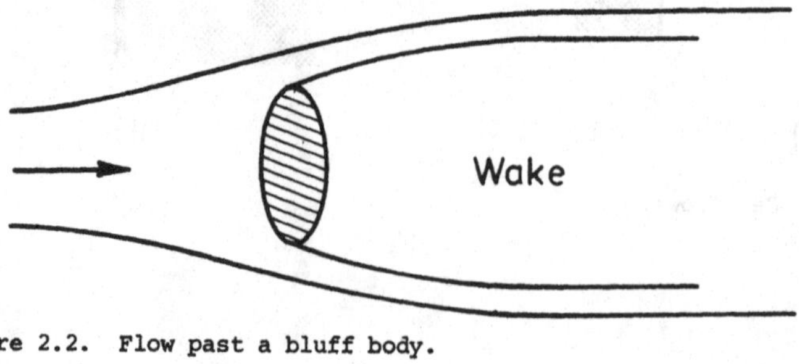

Figure 2.2. Flow past a bluff body.

2. GRAVITY WAVES

Probably the most interesting application of the irrotational flow model with free boundaries is the propagation of disturbances on the surface of an expanse of incompressible fluid in which gravity effects are important. This models the behaviour of the surface of the sea, and we can hope to find solutions which give results about tides and the motion of waves and other disturbances.

Before describing the model in detail, there are problems of notation we must consider. It will be very convenient to denote derivatives such as $\partial\phi/\partial x$ by the suffix form ϕ_x and to avoid the components x_i so that suffices on suffices do not appear. Two-dimensional problems are very common, using variables x and y with y in the vertical direction; z is also commonly used as the complex variable $x + iy$. Hence we shall use the variable s in the second horizontal direction to avoid ambiguity.

We consider an incompressible fluid unbounded in the x and s directions but contained in the vertical direction by a bottom defined by $y = -h(x,s,t)$. The free surface, on which the pressure is constant, is defined by $y = \eta(x,s,t)$, with $y = 0$ the mean surface height. We require motion at infinity to be bounded, and at $t = 0$ there is a prescribed flow. Then the boundary value problem for the velocity potential $\phi(x,y,s,t)$ and the unknown surface shape $\eta(x,s,t)$ can be written

$$\phi_{xx} + \phi_{ss} + \phi_{yy} = 0. \tag{2.3}$$

On $y = -h(x,s,t)$,

$$\phi_y = \frac{dy}{dt} = -\frac{dh}{dt} = -h_t - \phi_x h_x - \phi_s h_s. \tag{2.4}$$

On $y = \eta(x,s,t)$,

$$\left.\begin{aligned} &\phi_y = \eta_t + \phi_x\eta_x + \phi_s\eta_s, \\ &\text{and, using (1.17),} \\ &\phi_t + \frac{1}{2}\,|\text{grad}\,\phi|^2 + g\eta = 0, \end{aligned}\right\} \tag{2.5}$$

where the arbitrary function $G(t)$ in Bernoulli's equation is absorbed into ϕ_t. In addition there are conditions at infinity to be applied†,

---

†The system of equations (2.3), (2.4) and (2.5) is not sufficiently understood for us to be able to state the precise conditions at infinity necessary for a unique solution. By simplifying the general non-linear problem we shall obtain a model for which this difficulty can be resolved.

and ϕ, η and η_t are prescribed at $t = 0$. Without loss of generality we consider flows starting from rest so that at $t = 0$,

$$\phi = 0, \quad \eta = \eta_0(x,s), \quad \eta_t = 0. \tag{2.6}$$

The usual hodograph methods of the previous section are now not applicable even in the steady two-dimensional case because Bernoulli's equation no longer reduces to the simpler linear condition that q is constant on the free surface. It is not possible to find, or even prove the existence of, exact solutions to the full problem except in a few very special situations, and we are forced to investigate various limiting cases in which one or other of the parameters involved in the solution is small. There are essentially two non-dimensional parameters in the general gravity wave problem. One is ε, the ratio of the height or amplitude of the disturbance to the depth of the fluid, and the other is δ, the ratio of the depth to the horizontal scale or wavelength of the disturbance.

If $\varepsilon \ll 1$ we have *small disturbance theory* and if $\delta \ll 1$ we have long wave length or *shallow water theory*. Neither theory will be uniformly valid for all x,s,t, but we delay a discussion of the formal asymptotic expansions in either ε or δ of the full non-dimensional boundary value problem to Section III.3. In that section we also discuss possible flow situations when both ε and δ are small. We now continue with heuristic derivations of the equations of small disturbance theory and shallow water theory and develop their solutions.

3. SMALL DISTURBANCE THEORY

If the height of the surface disturbance $y = \eta$ above the mean height $y = 0$ is small compared to the mean depth, then the resulting flow will consist of small disturbances about a uniform state. These disturbances are called *Stokes waves*. Since the problem is invariant with respect to axes moving with constant speed, we need only consider this uniform state to be that of rest and for the problem of small disturbances on a moving stream take axes moving with the stream. Hence we can linearize the boundary conditions (2.5) on the free surface to give, as a first approximation,

$$\text{on } y = \eta(x,s,t), \quad \phi_y = \eta_t \quad \text{and} \quad \phi_t + g\eta = 0. \tag{2.7}$$

This can be further simplified by expanding ϕ_y and ϕ_t as power series in η and assuming that they have bounded derivatives, thus reducing the boundary condition onto $y = 0$. Hence (2.7) becomes

$$\text{on } y = 0, \quad \phi_{tt} + g\phi_y = 0. \tag{2.8}$$

If in addition the bottom is horizontal or the spatial variation in depth of the bottom is small, then (2.4) becomes

$$\text{on } y = -h_0(t), \quad \phi_y = -h_t, \tag{2.9}$$

where $h = h_0(t)$ + smaller terms.

This is now a linear boundary value problem for ϕ, and hence η, and the general solution can be obtained by the use of Fourier transforms in the x and s variables. However, we shall first give a more elementary discussion and look for certain periodic solutions of a two dimensional nature, that is, independent of s.

Wave Dispersion

A *travelling* or *progressive* wave is a disturbance of the form $\eta = f(x-ct)$, where f is a given function and c is the wave speed. In some physical systems, such as transverse waves on strings, it is possible to propagate a travelling wave of arbitrary profile f, and the wave speed is independent of f. Stokes waves on the surface of a fluid are not of this kind and it is only possible to propagate a strictly harmonic small amplitude wave without change of shape. We therefore look for a solution in the form $\eta = Ae^{ik(x-ct)}$, where for convenience we imply that the real part of all complex expressions is to be taken and k is real and positive. $|A|$ is then the amplitude of the wave, the wave length is $2\pi/k$ and k is the wave number. For a problem with a fixed horizontal bottom $y = -h_0$ we now look for a solution for ϕ in the form

$$\phi = B \cosh k\,(y+h_0)e^{ik(x-ct)}. \tag{2.10}$$

This satisfies Laplace's equation and the condition on the bottom. From the free surface condition (2.8) we require

$$c^2 = \frac{g}{k} \tanh kh_0, \tag{2.11}$$

and B is related to A by (2.7). Thus the wave speed c depends on the wave number k. It is easily verified that only a harmonic disturbance of this type can be propagated without change of shape. Such waves are called *dispersive*, and (2.11) is their *dispersion relation*. The relationship between c and k is sketched in Figure 2.3. If the water is very deep we can let $h \to \infty$, and then $\phi = Be^{ky+ik(x-ct)}$, where $c^2 = g/k$.

It is possible to observe these periodic waves on the surface of large expanses of water and to verify relation (2.11) experimentally. There is no net motion of the fluid, and the particle paths can be shown to be approximately ellipses with the major axes horizontal. For very deep water the particle paths are circles.

The rate at which energy is transmitted across a fixed vertical surface is the sum of the flux of the kinetic and potential energy of the fluid together with the rate at which work is done by the pressure forces. Thus the energy flux is

$$\int (\tfrac{1}{2}\rho q^2 + \rho gy)q\cdot ds + \int pq\cdot ds = \int_{-h}^{\eta} -\rho\phi_t\phi_x dy$$

on using (1.17). For the case of small amplitude waves on deep water, if $\phi = B_1 e^{ky} \cos k(x-ct)$, then the energy flux is

$$\int_{-\infty}^{0} \rho B_1^2 k^2 ce^{2ky} \sin^2 k(x-ct)dy = \tfrac{1}{2}\rho B_1^2 kc \sin^2 k(x-ct).$$

Hence the average flux of energy over a period is $\rho \frac{B_1^2}{4}\sqrt{kg}$. Now the energy stored per unit length is

$$\frac{1}{2}\rho\int_{-\infty}^{0} q^2 dy + \rho g\int_0^{\eta} ydy = \rho\frac{B_1^2 k}{4} + \rho\frac{B_1^2 k^2 c^2}{2g}\sin^2 k(x-ct).$$

The average energy stored per wave length is therefore $\rho\frac{B_1^2 k}{2}$. The rate at which energy is transmitted is the ratio of the average flux to the average energy stored and is therefore equal to $\frac{1}{2}(\frac{g}{k})^{\frac{1}{2}}$, that is, exactly half the wave speed.

A *forced* travelling wave on the surface, with wave speed not satisfying the dispersion relation (2.11), can be generated by a bottom whose shape is undergoing a wave motion or by a varying air pressure p_0. In the first case if the bottom has a profile $y = -h_0 + ae^{ik(x-ct)}$ where $|a| << h_0$, then the boundary condition on $y = -h_0$ is $\Phi_y = -ikcae^{ik(x-ct)}$. Then the possible forms for Φ and η are

$$\Phi = (A \cosh ky + B \sinh ky)e^{ik(x-ct)}$$

and

$$\eta = be^{ik(x-ct)}.$$

From the boundary condition on the bottom,

$$A \sinh kh_0 + B \cosh kh_0 = -ica,$$

and from (2.7), the conditions on the surface give

$$B = -icb \quad \text{and} \quad gb - ikcA = 0.$$

Thus

$$b = \frac{kc^2 a}{kc^2 \cosh kh_0 - g \sinh kh_0}, \tag{2.12}$$

and such a forced wave is possible provided that the denominator of b is bounded. That is, the speed and wave number k of the forced wave must not satisfy the free wave dispersion relation, even approximately. If they did *resonance* would occur and the linearized small disturbance theory would not be valid. If we change to axes moving with speed c these solutions give the flow of a uniform stream of speed c over a fixed bottom of harmonic shape. Since (2.11) will only have real roots for k when $c^2 < gh_0$, harmonic waves of wave number k can exist on a stream, in addition to any forced wave, provided that the speed of the stream is less than $\sqrt{gh_0}$. Such a stream is said to be *subcritical*, and $\sqrt{gh_0}$ is the critical stream speed.

Group Velocity

Consider now the propagation of a small disturbance which is not strictly harmonic, so that its profile changes in time. We would expect to be able to use Fourier analysis, that is, to write η as an integral over all possible wave numbers in the form

$$\eta = \int_{-\infty}^{\infty} A(k) e^{ik(x-c(k)t)} dk, \tag{2.13}$$

where ϕ has a corresponding form, $c(k)$ is given by (2.11), and $A(k)$ is such that the integral converges. At time $t = 0$, $\int_{-\infty}^{\infty} A(k) e^{ikx} dk = \eta_0(x)$ so that η_0 is the Fourier transform of $A(k)$ and

$$A(k) = \frac{1}{2\pi} \int_{-\infty}^{\infty} e^{-ikx} \eta_0(x) dx. \tag{2.14}$$

The integral will converge if $\eta_0(x)$ tends to zero for large enough $|x|$. The results (2.13) and (2.14) could have been obtained directly by using Fourier transforms on the boundary value problem (2.3), (2.8), (2.9). They give the profile shape in the form of a double integral which cannot in general be explicitly integrated. We can however evaluate the integral asymptotically for large time, and to do this we use the *method of stationary phase*. First we observe that, since each wave travels with finite speed, if t is large then so is x. We therefore examine the case when $x = Gt$, where G is a constant speed which is comparable in size with other velocity scales in the problem. The integral (2.13) then reduces to the form $\int_{-\infty}^{\infty} A(k)e^{ith(k)}dk$ where $h(k) = Gk - kc(k)$. For a detailed analysis of the asymptotic value of such integrals as $t \to \infty$ reference should be made to Copson [7], but we give here a very brief account of the method. The argument used is that when t is large the integrand oscillates rapidly with a slowly varying amplitude so that the positive and negative contributions from each cycle almost cancel each other out. This cancellation will be least effective when the oscillations are most widely spread. This will occur when $h(k)$ varies slowly with k, so that the main contribution to the integral will come from values of k which are the roots of $h'(k) = 0$. The method of stationary phase shows that the correct asymptotic form is obtained by expanding the integrand as an asymptotic series in $(k-k_0)$ where $h'(k_0) = 0$.†

For the given integral, the dominant wave number k_0 is the root of $\frac{\partial}{\partial k}(Gk - kc(k)) = 0$, or

$$G = \frac{\partial}{\partial k}(kc(k)). \tag{2.15}$$

For simplicity we only consider waves on a sea of infinite depth so that $c^2 = g/k$ and

$$k_0 = \frac{g}{4G^2} = \frac{gt^2}{4x^2}.$$

Now we write $k = k_0 + p$ and expand (2.13) in powers of p so that

$$\eta = \int_{-\infty}^{\infty} \{A(k_0) + O(p)\}\exp\left\{-\frac{igt}{4G} + \frac{ip^2G^3t}{g} + O(p^3)\right\}dp.$$

Putting $p' = p\,\frac{G^{3/2}t^{1/2}}{g^{1/2}}$, this becomes

$$\eta = A(k_0)\,\frac{g^{1/2}}{G^{3/2}t^{1/2}}\,e^{-igt/4G}\int_{-\infty}^{\infty} e^{ip'^2}\,dp' + \text{smaller terms}.$$

We can see that

$$\eta \sim \frac{Ce^{\frac{-igt}{4G}}}{t^{1/2}}, \tag{2.16}$$

where C is a constant. Thus for large values of x and t the wave

---

†This method is a special case of the method of steepest descents for complex integrals.

profile is oscillatory with an amplitude which decays like $t^{-1/2}$.

$G(k_0)$ as defined by (2.15) has a physical interpretation and is called the *group velocity*. If an observer moves with the group velocity relative to the waves he will see a profile which is dominated after a long time by waves whose wave number is k_0. Thus, after a long time, waves of any given wave number k will travel with their wave speed $c(k)$, called their phase speed, only if no other harmonics are present in the initial disturbance to interact with them and cause them to disperse.

An alternative illustration is to consider a fixed observer viewing waves propagating from some general disturbance which occurred a long time earlier a long distance away. The observer will see a harmonic wave train whose wave number depends through (2.15) on the ratio G of the distance from the initial disturbance to the time which has elapsed since it occurred, so that as time passes G will slowly decrease. From (2.11) we see that c and G are decreasing functions of k for Stokes waves, as shown in Figure 2.3. Thus in this case the stationary observer will see waves of gradually decreasing wavelength as time goes by. For very deep water, $G = \frac{1}{2}\sqrt{g/k} = \frac{1}{2}c$, which has been shown to be the speed at which energy is transmitted by the wave. This is true generally, and an alternative definition of group velocity is that it is the speed with which the energy of the wave is transmitted.

An alternative derivation of the long time solution to an initial value problem of this kind proceeds by a direct approach to the equations without using Fourier Transforms. Since both time t and the x scale are large, we rescale them with an artificial parameter λ such that

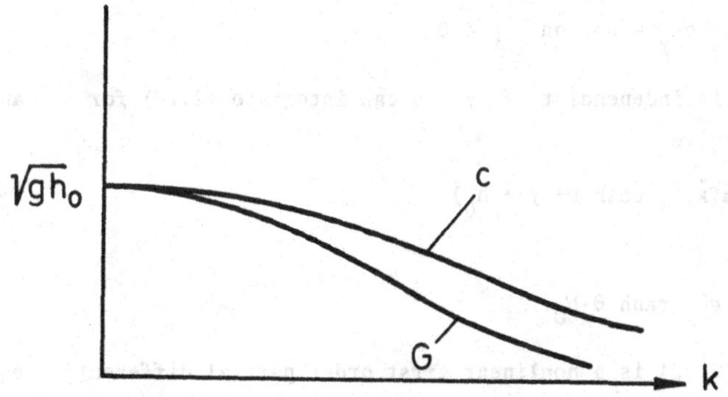

Figure 2.3. Dispersion relation for Stokes waves

$\lambda >> 1$. Then, putting $x = \lambda\bar{x}$ and $t = \lambda\bar{t}$, the boundary value problem becomes

$$\phi_{\bar{x}\bar{x}} + \lambda^2\phi_{yy} = 0 \quad \text{for} \quad -h_0 < y < 0, \tag{2.17}$$

$$\phi_y = 0 \quad \text{on} \quad y = -h_0, \tag{2.18}$$

and

$$\phi_{\bar{t}\bar{t}} + \lambda^2 g\phi_y = 0 \quad \text{on} \quad y = 0. \tag{2.19}$$

We now look for a solution of the type used in the W.K.B. method for ordinary differential equations† and write

$$\phi = A(\bar{x},y,\bar{t})e^{i\lambda\theta(\bar{x},y,\bar{t})}.$$

Then

$$\phi_{\bar{x}} = i\lambda\theta_{\bar{x}}\phi + O(1),$$

$$\phi_{\bar{x}\bar{x}} = -\lambda^2\theta_{\bar{x}}^2\phi + O(\lambda),$$

and similar expressions hold for $\phi_{\bar{t}}$ and $\phi_{\bar{t}\bar{t}}$. Substituting in (2.17) and equating powers of λ^4 gives

$$\theta_y^2 = 0,$$

so that θ depends only on $\bar{x}$ and $\bar{t}$. Then for the $O(\lambda^2)$ term in (2.17) we get

$$A_{yy} - \theta_{\bar{x}}^2 A = 0 \quad \text{for} \quad -h_0 < y < 0, \tag{2.20}$$

and the boundary conditions (2.18) and (2.19) become

$$A_y = 0 \quad \text{on} \quad y = -h_0$$

and $$\tag{2.21}$$

$$-\theta_{\bar{t}}^2 A + gA_y = 0 \quad \text{on} \quad y = 0.$$

Since θ is independent of y we can integrate (2.20) for A and use (2.21) to give

$$A = \alpha(\bar{x},\bar{t}) \cosh \theta_{\bar{x}}(y + h_0)$$

where

$$\theta_{\bar{t}}^2 = g\theta_{\bar{x}} \tanh \theta_{\bar{x}}h_0. \tag{2.22}$$

Equation (2.22) is a nonlinear first order partial differential equation

---

†See Carrier and Pearson [4].

for the *phase function* $\theta(\bar{x},\bar{t})$ and may be solved by the use of Charpit's equations along its characteristics.[†] However, a simpler approach is possible for this particular equation since $\bar{x},\bar{t}$ and θ do not appear explicitly.

We define $k = \theta_{\bar{x}}$ and $\omega = -\theta_{\bar{t}}$ so that (2.22) gives a relation $\omega = \omega(k) = (gk \tanh kh)^{1/2}$, which is exactly the same as equation (2.11). This is not unexpected since this approach is equivalent to regarding θ as a linear function of x and t. Eliminating θ, we see that ω and k also satisfy the equation $\partial\omega/\partial\bar{x} + \partial k/\partial\bar{t} = 0$[*], and since ω is a function of k this equation can be rewritten as $\partial k/\partial\bar{t} + d\omega/dk\,(\partial k/\partial\bar{x}) = 0$. This is a quasilinear equation for k which has the general solution $k = F(\bar{x} - \frac{d\omega}{dk}\bar{t})$ for an arbitrary function F. This may be interpreted as k = constant on the lines $x - \frac{d\omega}{dk}t$ = constant. These lines, or rays, are propagated with speed $d\omega/dk$, which is the group velocity. On such lines, both $\theta_{\bar{x}}$ and $\theta_{\bar{t}}$ are constant and so $\theta = k\bar{x} - \omega\bar{t}$, or $\lambda\theta = kx - \omega t$. Thus for large times, waves with wave number k will dominate if we travel with the group velocity of these waves. This approach can be used in any dispersive problem and the dispersion relation $\omega = \omega(k)$ is then converted into a first order nonlinear partial differential equation $-\theta_{\bar{t}} = \omega(\theta_{\bar{x}})$. If this relation contains x or t explicitly (which is possible) then the equation will not have a simple integral of the type found for (2.22), and the concept of group velocity requires a more careful explanation.

Example 1. Stokes waves with surface tension.

If we introduce surface tension effects into the problem, the linearized boundary conditions (2.7) on the surface $y = 0$ become

$$\phi_y = \eta_t \quad \text{and} \quad \rho\phi_t + \rho g\eta = T\eta_{xx}$$

(provided the free boundary slope remains small), where T is the surface tension per unit length. Eliminating η gives

$$\phi_{tt} + g\phi_y = \frac{T}{\rho}\phi_{yxx} \quad \text{on} \quad y = 0$$

as the boundary condition on the surface, and the dispersion relation for a wave of the form (2.10) is

$$\omega^2 = k^2c^2 = \left(gk + \frac{Tk^3}{\rho}\right)\tanh kh_0.$$

† See Chester [5] for details of this method.

* This equation may be thought of as a statement of conservation of the number of waves in a wave train.

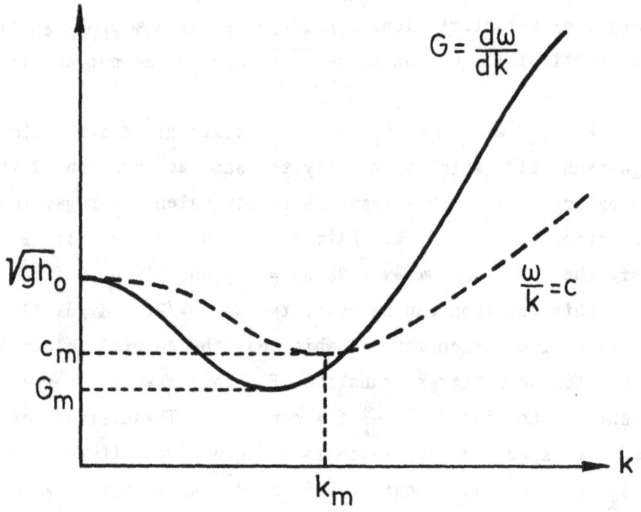

Figure 2.4. Dispersion relation for Stokes waves with surface tension.

If $T \ll \rho g h_0^2$, then as $k \to 0$,

$$\frac{d\omega}{dk} \sim \sqrt{gh_0}\,(1 - \tfrac{1}{2}k^2h_0^2 + \ldots),$$

whereas for $k \to \infty$,

$$\frac{d\omega}{dk} \sim \frac{3}{2}\left[\frac{Tk}{\rho}\right]^{1/2}.$$

Then we can see from Figure 2.4 that $G = d\omega/dk$ has a minimum value G_m. The graph for $\omega/k = c$ is drawn on the same diagram; it shows that there is also a minimum value c_m of c which occurs at $k = k_m$. This is in contrast to the case where $T = 0$ which is shown in Figure 2.3.

The effect of surface tension on the wave is negligible for small k but is the dominant effect when k is large. There are thus two families of waves depending on whether k is greater or less than k_m, and for values of c between c_m and $\sqrt{gh_0}$ there are two possible waves travelling with the same speed c. The waves with $k > k_m$ are called *capillary waves*. For all waves the wave speed must exceed c_m and the group velocity must exceed G_m. If the initial disturbance is over a finite domain near $x = 0$ then for large times there will be no disturbance for $x < G_m t$, and we see that the effect of surface tension is to restrict the extent to which the disturbance spreads out with time.

Example 2. Waves at an interface

Another example is to consider waves at an interface between two immiscible fluids of different densities, ρ and ρ', contained between $y = -h_0$ and $y = h_0'$. If the mean position of the interface is at $y = 0$, the linearized boundary conditions at the interface $y = \eta(x,t)$ reduce to

$$\rho(g\eta + \phi_t) = \rho'(g\eta + \phi_t')$$

and

$$\eta_t = \phi_y = \phi_y' \quad \text{on} \quad y = 0,$$

where ϕ and ϕ' are the velocity potentials in the two fluids. The dispersion relation for a harmonic wave on the interface is therefore

$$kc^2 = \frac{(\rho-\rho')g}{\rho \coth kh_0 + \rho' \coth kh_0'},$$

which is only possible if $\rho > \rho'$. This is the obvious physical condition that the heavier fluid should be underneath in a stable situation.

4. THREE DIMENSIONAL STOKES WAVES

Perhaps the simplest three dimensional situation is that with cylindrical symmetry. The appropriate velocity potential which is periodic in time of the form e^{-ikct} is

$$\phi = BJ_0(kr) \cosh k(y+h_0)e^{-ikct}, \tag{2.23}$$

where c is defined by (2.11) as before, $r^2 = x^2 + s^2$, and J_0 is the zeroth order Bessel function. This is called a circular wave, and a simple harmonic profile in this case will not propagate without change of shape. We can sum over all wave numbers to obtain for a general disturbance with cylindrical symmetry

$$\eta(r,t) = \int_{-\infty}^{\infty} A(k)J_0(kr)e^{-ikct}dk. \tag{2.24}$$

To complete the determination of $A(k)$ from the initial conditions we need an inversion formula for the Fourier-Bessel integral. The result is

$$\begin{aligned} A(k) &= k\int_0^{\infty} \eta_0(r)J_0(kr)rdr \quad \text{for} \quad k > 0, \\ &= 0 \quad\quad\quad\quad\quad\quad\quad\quad\quad \text{for} \quad k < 0. \end{aligned} \tag{2.25}$$

The convergence of this integral is assured if $\eta_0(r)$ tends to zero fast enough as $r \to \infty$. It is a complicated matter to evaluate (2.24) even asymptotically for large t and we shall not do so here.

The WKB expansion method may also be used in a three dimensional situation to study waves a long time after their initiation. As in the last section we introduce a large parameter λ and put $x = \lambda\bar{x}$, $s = \lambda\bar{s}$, $t = \lambda\bar{t}$ and $\phi = A(\bar{x},y,\bar{s},\bar{t})e^{i\lambda\theta(\bar{x},\bar{s},t)}$. By writing $\omega = -\theta_{\bar{t}}$, $\ell = \theta_{\bar{x}}$ and $m = \theta_{\bar{s}}$, we obtain the dispersion relation $\omega = \omega(\ell,m)$. For a steady problem this reduces to

$$\omega(\ell,m) = 0 \quad \text{and} \quad \frac{\partial \ell}{\partial \bar{s}} = \frac{\partial m}{\partial \bar{x}} .$$

Rewriting the relation between ℓ and m in the form $\ell = \ell(m)$ we can argue, just as before, that ℓ and m are constant on the lines $ds/dx = -d\ell/dm$. Hence the waves will propagate in a direction $(-d\ell/dm)$ and the concept of group velocity becomes 'group direction'.

As an example, we consider a point source moving with constant speed U in the negative x direction on water of infinite depth. This may be regarded as a simple model for a ship travelling over a large expanse of deep water. Relative to axes moving with the ship this is a steady problem, and the boundary condition on $y = 0$ is obtained from (2.8) by writing $-U\partial/\partial x$ in place of $\partial/\partial t$. Then for a single wave train

$$\phi = Ae^{-ky+i(\ell x+ms)},$$

where $k^2 = \ell^2 + m^2$, and $U^2\ell^2 = gk$ is the dispersion relation. To evaluate $G = d\ell/dm$ we use $U^4\ell^4 = g^2(\ell^2+m^2)$ to give

$$G = \frac{(\alpha^2\ell^2-1)^{\frac{1}{2}}}{2\alpha^2\ell^2 - 1},$$

where $\alpha = U^2/g$. Hence G has a maximum value $G_m = \frac{1}{2\sqrt{2}}$ which occurs when $\ell^2 = 3/2\alpha^2$ and $m^2 = 3/4\alpha^2$, as shown in Figure 2.5. Thus all the waves are contained within a wedge of semiangle $\tan^{-1}(1/2\sqrt{2}) = \sin^{-1}(1/3)$, or about $19\frac{1}{2}°$. This is shown in Figure 2.6. At any angle less than this limiting angle, two waves occur, as can be seen from Figure 2.5. The solution is singular on the limiting line and the amplitude is larger there. This is the *bow wave* of a ship and is readily observed in practice. This problem is discussed more fully by Stoker [24].

In the two unsteady examples (2.13) and (2.24) considered above we have been able to obtain a unique solution on the assumption that the disturbance tends to zero at infinity. In more general problems involving the propagation of wave trains to infinity a weaker condition may be

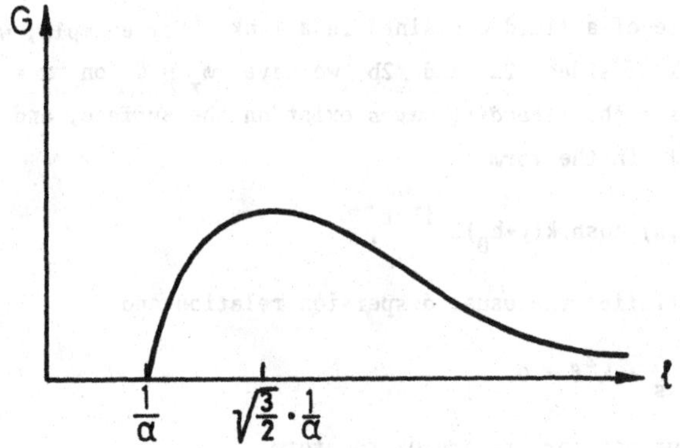

Figure 2.5. Dispersion relation for ship waves.

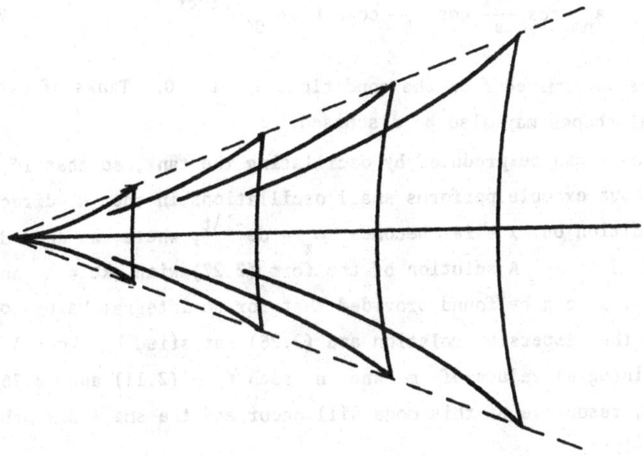

Figure 2.6. Ship waves; wave crests and bow wave.

necessary for any solution to exist at all. This is called the *radiation condition* and expresses the physical idea of an outgoing wave [see Exercise 9].

A further extension of small disturbance theory is to consider waves on the surface of a fluid contained in a tank. For example, with a rectangular tank of sides 2a and 2b we have $\phi_x = 0$ on $x = \pm a$ and $\phi_s = 0$ on $s = \pm b$. *Standing waves* exist on the surface, and we look for a solution ϕ in the form

$$\phi = f(x,s)\cosh k(y+h_0)e^{-ikct},$$

where c satisfies the usual dispersion relation and

$$f_{xx} + f_{ss} + k^2 f = 0.$$

Possible solutions for f are of the form

$$f = a_{nm}\cos\frac{n\pi x}{a}\cos\frac{m\pi s}{b},$$

where a_{nm} is an arbitrary constant and n,m are integers such that

$$\pi^2\left(\frac{n^2}{a^2} + \frac{m^2}{b^2}\right) = k^2. \tag{2.26}$$

The general solution is therefore

$$\phi = \sum_{n=1}^{\infty}\sum_{m=1}^{\infty} a_{nm}\cos\frac{n\pi x}{a}\cos\frac{m\pi s}{b}\cosh k(y+h_0)e^{-ikct}, \tag{2.27}$$

where a_{nm} are determined from the conditions at $t = 0$. Tanks of other cross-sectional shapes may also be discussed.

Forced waves can be produced by oscillating the tank, so that if the tank in the above example performs small oscillations in the x direction, the condition on $x = \pm a$ becomes $\phi_x = de^{-i\lambda t}$, where λ and d are given and $d \ll a$. A solution of the form (2.27) with $kc = \lambda$ and c given by (2.11) can be found provided that for no integral values of n or m are the dispersion relation and (2.26) satisfied by $kc = \lambda$. If there are integral values of m and n such that (2.11) and (2.26) are satisfied, resonance of this mode will occur and the small disturbance theory is invalid.

5. TIDAL WAVES

So far we have only considered waves which constitute a small disturbance on relatively deep water. If the depth of the water is comparable with the height of the wave, it is still possible to derive a linear theory provided that the depth is small compared with the wavelength. The key fact is that gravity is the dominant effect in the vertical direction so that the pressure is hydrostatic; this follows from the assumption that the scale for y is small compared to the length scale for x. If the surface is given by $y = \eta(x,t)$, and atmospheric pressure is a constant p_0, this leads to

$$p = -\rho g y + \rho g \eta + p_0. \tag{2.28}$$

The next assumption necessary for a linear theory is to assume that the magnitude of the horizontal velocity u is small compared with the critical speed $\sqrt{gh}$, where h is the mean depth of the water. This assumption allows us to linearize the x component of the momentum equation (1.8) to obtain

$$u_t = -\frac{1}{\rho} p_x = -g\eta_x, \tag{2.29}$$

on using (2.28). From (2.29) $\frac{\partial}{\partial t}(u_y) = 0$, and hence if $u_y = 0$ initially then $u_y = 0$ for all t and we can integrate the continuity equation

$$\int_{-h}^{\eta} (u_x + v_y)dy = 0$$

to obtain

$$(\eta+h)u_x = -v(x,\eta,t).$$

Now $v(x,\eta,t) = \partial\eta/\partial t$ from (2.7), and linearizing again we have

$$hu_x = -\eta_t. \tag{2.30}$$

Finally, equations (2.29) and (2.30) lead to the wave equation

$$\frac{\partial}{\partial x}\left(gh \frac{\partial u}{\partial x}\right) = \frac{\partial^2 u}{\partial t^2}, \tag{2.31}$$

and when h is constant waves will travel with speed $\sqrt{gh}$. The interesting case when the nonlinear terms are not negligible will be considered under the heading Shallow Water Theory in the next chapter.

The steady flow in a straight horizontal channel of slowly varying cross-section may also be analyzed by regarding the flow as being virtually one dimensional and the pressure hydrostatic. We suppose that the bottom is horizontal but that the sides of the channel are such that its breadth $b(x)$ is a slowly varying function of x. Let the depth of the water be h and the horizontal velocity be u. Then from continuity

$$ubh = \text{constant} = Q, \tag{2.32}$$

and from Bernoulli's equation (1.17)

$$\tfrac{1}{2}u^2 + gh = \text{constant} = gH, \tag{2.33}$$

where H is the pressure head. It is convenient to introduce the *Froude Number* $F = u/\sqrt{gh}$ so that equations (2.32) and (2.33) can be rewritten as

$$h = \frac{H}{1 + \frac{1}{2}F^2} \quad \text{and} \quad b = \frac{Q(1+\frac{1}{2}F^2)^{3/2}}{(gH^3)^{1/2}F}. \tag{2.34}$$

It is interesting to consider flow in a channel which first converges and then diverges. From the graph of b in Figure 2.7 we see that it is possible to have a flow through such a channel in which the flow is subcritical $(F < 1)$ in the converging part of the channel and then switches to supercritical $(F > 1)$ in the divergent part. An exactly similar phenomenon occurs for the one dimensional flow of a gas through a pipe of slowly varying cross-section; this is described in Chapter IV, Section 2.

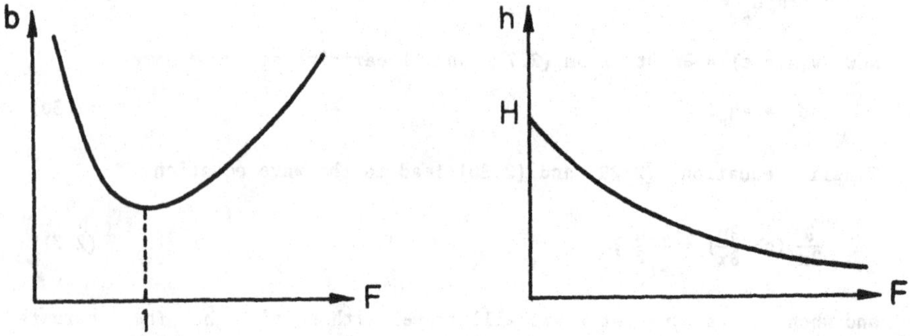

Figure 2.7. Variation of flow variables with Froude number.

EXERCISES

1. An inviscid stream of depth h flows with speed U over a horizontal flat bottom. Obtain the dispersion relation for the propagation of harmonic waves on the surface. Deduce that standing waves are possible only if $U^2 < gh$. If the bottom is given by $y = -h + a \cos kx$ ($a \ll h$), obtain an expression for the shape of the free surface if $U^2 > gh$.

2. Show that the particle paths for Stokes waves on still water of depth h are ellipses.

3. The pressure in the atmosphere above a two dimensional inviscid stream with velocity c and depth h_0 is $p_0 + a \cos mx$, where $a \ll p_0$. Find the shape of the free surface. How will surface tension affect this result? Obtain a condition which must hold for surface tension effects to be important. If surface tension is negligible, under what conditions is resonance possible?

4. Show that the group velocity for Stokes waves on water of depth h_0 is the same as the velocity with which energy is propagated by the waves.

5. Obtain the dispersion relation for waves at an interface between two liquids of densities ρ and ρ' and mean depths h and h' when surface tension effects are important. If h and h' are large, show that there is a minimum value c_0 for the wave speed c and that for $c > c_0$ there are two waves with the same wave speed.

6. A rectangular tank with its bottom horizontal and given by $y = 0$ has its side walls at $x = \pm a$, $s = 0,b$ and is filled with water to depth h. Initially the water is at rest under an external pressure $p_0 - p_1\, x/a$ applied to its free surface. This pressure is suddenly reduced to p_0 at $t = 0$. On the assumption that $p_1/\rho gh,\ p_1/\rho ga \ll 1$, find the form of the free surface for $t \geq 0$.

7. A two dimensional tank with vertical sides $x = \pm d$ and horizontal bottom $y = -h$ is filled with inviscid incompressible fluid to mean depth h. At $t = 0$ the tank starts to vibrate horizontally so that the vertical sides are given by $x = \pm d + \varepsilon d \sin \lambda t$, where $\varepsilon \ll h/d \sim 1$. Show that the free surface $y = \eta(x,t)$ is given by

$$\eta = \varepsilon\lambda \frac{\sin \lambda t}{g}\left[\lambda dx + \sum_{n=0}^{\infty} a_n \cos \frac{n\pi(x+d)}{2d} \cosh \frac{n\pi h}{2d}\right],$$

where the a_n are suitably defined. Discuss the possibility of resonance.

8. Fluid of depth h is contained in a square tank with vertical sides at $x = 0, a$ and $s = 0, a$. Determine the possible frequencies of standing waves on the surface. Show that a mode of oscillation is possible in which there is no vertical motion in the diagonal plane $x = s$ and in which $\omega^2 = \sqrt{5}\,\frac{g\pi}{a}\tanh(\sqrt{5}\,\frac{\pi h}{a})$. Show that a fluid particle in this plane is moving normally to the plane with a speed proportional to $\cosh k(y+h)\cos\omega t\sin\frac{\pi x}{a}\cos\frac{2\pi x}{a}$, where $k = \sqrt{5}\,\frac{\pi}{a}$.

9. Incompressible fluid is contained at rest in $-\infty < y < 0$, $a < r < \infty$, where r is measured radially from the vertical y axis. A wave motion is produced by oscillating the boundary so that $a = a_0 + \varepsilon e^{-ikct+ky}$ where $kc^2 = g$ and $\varepsilon \ll a_0$. Show that the velocity potential

$$\phi = \varepsilon f(kr)e^{-ikct+ky},$$

where f satisfies

$$f'' + \frac{1}{kr}f' + f = 0; \quad f'(ka_0) = -ic; \quad f'(\infty) - if(\infty) = 0.$$

What is the physical interpretation of the condition applied as $r \to \infty$?

10. Water flows steadily from a large reservoir into a straight open channel of constant breadth b_1, passing first through a convergent-divergent section in which the minimum breadth is $b_m < b_1$. The bottom of the channel is horizontal. The water in the reservoir is of depth H. The stream enters the final straight part of the channel with Froude number $F = u/\sqrt{gh} = 2$. Assuming the channel width varies slowly, show that $b_1/b_m = \sqrt{2}$ and that the rate of flow is $Q = 2b(gH^3/27)^{\frac{1}{2}}$.

Chapter III
Nonlinear Surface Waves

1. SHALLOW WATER THEORY

We now consider the situation in which the mean depth h_0 of the fluid is comparable with the amplitude of the surface waves but small compared with the lateral scale or wavelength λ of the disturbance. The same assumption was used in discussing Tidal Waves, but in this case we shall not assume that the disturbance is small, and the problem will be essentially nonlinear.

To see the implications of one assumption we nondimensionalize the variables by writing $x = \lambda\hat{x}$, $y = h_0\hat{y}$, $u = U\hat{u}$ and $t = (\lambda/U)\hat{t}$, where U is a typical horizontal velocity. From the continuity equation (1.7), the correct nondimensionalization for v is $v = (Uh_0/\lambda)\hat{v}$. Similarly, after considering the x component of the momentum equation (1.8) we write $p = \rho U^2\hat{p}$. Then from the y component of this equation

$$\frac{\partial\hat{p}}{\partial\hat{y}} = -\frac{gh_0}{U^2} + O\left(\frac{h_0^2}{\lambda^2}\right),$$

and it is clear that $h_0 << \lambda$ implies that the inertia terms can be neglected in this equation and the pressure is hydrostatic. Reverting to dimensional variables,

$$p = -\rho gy + \rho g\eta, \tag{3.1}$$

where $y = \eta(x,t)$ is the equation of the surface. Substituting for p from (3.1) in the x component of the momentum equation gives

$$\frac{du}{dt} = -g\,\frac{\partial\eta}{\partial x}, \tag{3.2}$$

so that the convective derivative of u is independent of y. If we now assume that u is independent of y initially then u will remain independent of y for all t. Thus the problem has become one dimensional in space and (3.2) may be written

$$\frac{\partial u}{\partial t} + u\frac{\partial u}{\partial x} + g\frac{\partial \eta}{\partial x} = 0. \tag{3.3}$$

We assume that the bottom of the channel is horizontal and consider conservation of mass for the fluid contained between x and $x + \delta x$ to obtain the second equation

$$\frac{\partial \eta}{\partial t} + \frac{\partial}{\partial x}(u\eta) = 0. \tag{3.4}$$

This equation could also be obtained by integrating the continuity equation (1.6) with respect to y as was done in the section on Tidal Waves.

Equations (3.3) and (3.4) are thus two nonlinear equations for u and η. It is convenient to write $s^2 = g\eta$ and add and subtract equations (3.3) and (3.4) to derive equations in the form

$$\left[\frac{\partial}{\partial t} + (u \pm s)\frac{\partial}{\partial x}\right](u \pm 2s) = 0. \tag{3.5}$$

The form (3.5) shows that along certain 'lines' in the x,t plane defined by $dx/dt = u \pm s$, we have the integrals $u \pm 2s$ = constant. These lines are called the *characteristics,* and the integrals $u \pm 2s$ are the *Riemann invariants,* both of which clearly exist for all values of u and s. This implies that the system (3.3) and (3.4) is everywhere *hyperbolic*†. The existence of two families of characteristics greatly simplifies the theory and numerical solution of hyperbolic systems. Reference may be made to Courant and Hilbert [8] for details of the general theory. If the problem had been linear, the characteristics could have been found by a simple integration of a first order ordinary differential equation, and then a knowledge of the integrals along the characteristics would complete the solution. However, in a nonlinear problem the characteristic slope depends on the unknown dependent variables and cannot be found by integration. In certain special geometries, and with special boundary conditions, one family of characteristics reduces to a set of straight lines and then an analytical solution is possible. This is called a *simple wave flow.* We now consider several examples of such flows.

---

†See Appendix for results on hyperbolic partial differential equations.

Example 1. Dam break problem.

Consider a model for a dam in which fluid of depth h_0 is contained at rest in $x < 0$. At time $t = 0$ the containing wall at $x = 0$ is removed and the fluid flows into the constant pressure region $x > 0, y > 0$. Figure 3.1 shows the wave profile for $t < 0$ and $t > 0$ and gives the characteristic diagram in the x,t plane.

The characteristics in $t < 0, x < 0$ are, from (3.5), given by $dx/dt = \pm s_0$ since in this region $u = 0$, $s = s_0 = \sqrt{gh_0}$. The positive (negative) characteristics in this region are therefore straight lines $x = s_0 t + \text{const}$ ($x = -s_0 t + \text{const}$). Consider a point P in $t > 0$, $x < 0$ such that it is the intersection of a positive and negative characteristic, both of which intersect the x axis in $x < 0$. On the positive characteristic through P, which might be a curved line in $t > 0$, we have $u + 2s = 2s_0$ and on the negative characteristic through P, $u - 2s = -2s_0$. Thus, at P, $u = 0$ and $s = s_0$ so that P lies in the undisturbed region, and the characteristics through P are straight lines with slope $\pm s_0$. This undisturbed region, called a *zone of silence*, is bounded by $x = -s_0 t$, the extreme negative characteristic which intersects the negative x axis. Thus the disturbance initiated at time $t = 0$ at $x = 0$ is propagated into the water with the disturbance or wave front A having a speed s_0.

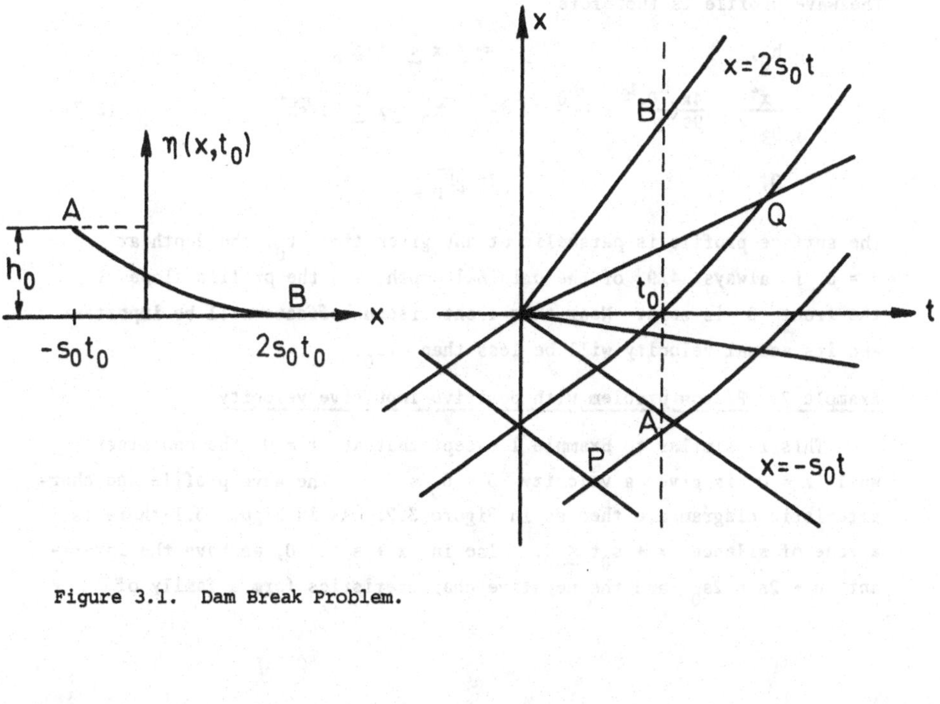

Figure 3.1. Dam Break Problem.

Any point Q in $x + s_0 t > 0$ will lie on a positive characteristic which emanates from the undisturbed region. Hence along this characteristic $u + 2s = 2s_0$, and this relation will therefore hold everywhere and is an invariant for any continuous flow. On the negative characteristic through Q, $u - 2s$ is a constant so that u and s must be constant on every negative characteristic, but their constant values vary from one negative characteristic to another. The family of negative characteristics must therefore be straight lines, and except at the origin only one negative characteristic can pass through any given point Q. Hence they must form a centred fan as in Figure 3.1, bounded by the negative characteristic corresponding to $s = 0$, that is, the wave front denoted by B in Figure 3.1. On this line in the x,t plane, $s = 0$, $u = 2s_0$ from the invariant $u + 2s = 2s_0$, and its slope is $2s_0$. The region of flow $-s_0 t < x < 2s_0 t$ is called a *centred simple wave*, and the solution can be obtained by simple geometry from the x,t plane as follows. The slope of a negative characteristic is

$$\frac{dx}{dt} = u - s = \frac{x}{t},$$

and using $u+2s = 2s_0$, the solution is

$$s = \frac{1}{3}(2s_0 - \frac{x}{t}) \quad \text{and} \quad u = \frac{1}{3}(2s_0 + 2\,\frac{x}{t}). \tag{3.6}$$

The wave profile is therefore

$$\begin{aligned} \eta &= h_0, & -\infty \le x \le -t\sqrt{gh_0} \\ &= \frac{x^2}{9gt^2} - \frac{4x}{9t}\left(\frac{h_0}{g}\right)^{\frac{1}{2}} + \frac{4h_0}{9}, & -t\sqrt{gh_0} \le x \le 2t\sqrt{gh_0} \\ &= 0, & 2t\sqrt{gh_0} \le x \le \infty. \end{aligned} \tag{3.7}$$

The surface profile is parabolic at any given time t_0, the depth at $x = 0$ is always 4/9 of the original depth, and the profile slope at the front B is zero. Near this front viscous effects will be important and its actual velocity will be less than $2s_0$.

Example 2. Piston problem with positive impulsive velocity.

This is similar to Example 1 except that at $t = 0$ the container wall $x = 0$ is given a velocity $0 < U_0 < 2s_0$. The wave profile and characteristic diagram are then as in Figure 3.2. As in Figure 3.1 there is a zone of silence $x + s_0 t \le 0$. Also in $x + s_0 t > 0$, we have the invariant $u + 2s = 2s_0$, and the negative characteristics form a family of

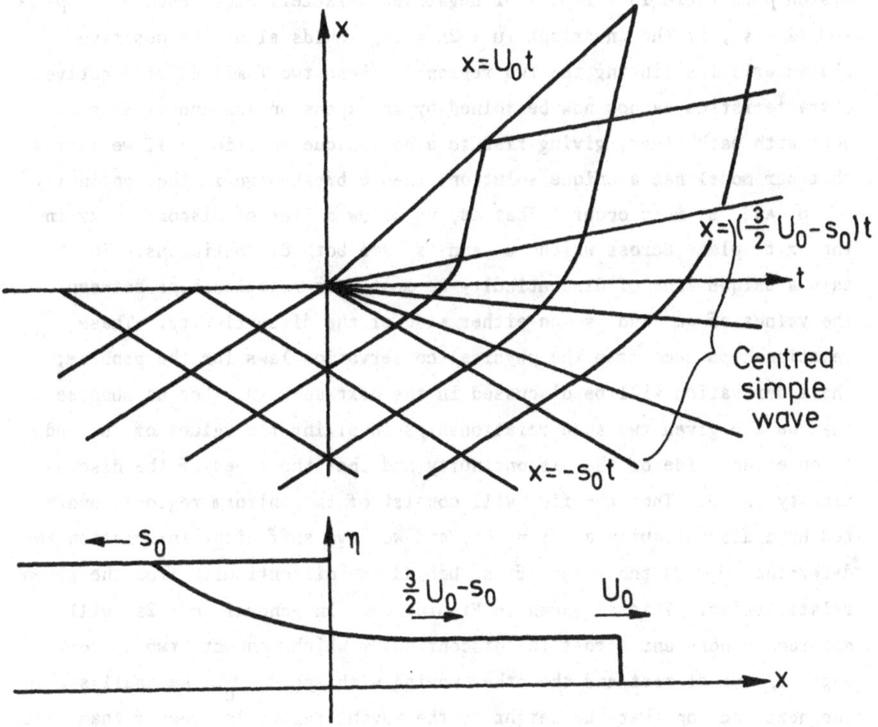

Figure 3.2. Impulsive piston problem; expansion case.

straight lines. Part of this family will form a fan through the origin and the remainder will intersect the 'piston curve' $x = U_0t$. On the piston curve, $u = U_0$ and hence $s = s_0 - U_0/2$. Thus the slope of every negative characteristic which intersects the piston curve is $U_0 - (s_0 - U_0/2) = 3U_0/2 - s_0$, and they form a family of parallel straight lines bounded by $x = (3U_0/2 - s_0)t$. The region $x \geq (3U_0/2 - s_0)t$ is therefore a uniform region in which $u = U_0$, $s = s_0 - U_0/2$. In the region $-s_0t < x < (3U_0/2 - s_0)t$ joining the two uniform regions there is a centred simple wave, and the solution is given by (3.6). If $U_0 > 2s_0$ the container wall separates from the fluid and the problem reduces to Example 1.

Example 3. Piston problem with negative impulsive velocity.

The container wall is now moved impulsively with speed U_0 towards the fluid. Associated with the region of fluid at rest there is a family of negative characteristics, each of slope $-s_0$, and emanating from the piston path there is a family of negative characteristics, each of slope $-3U_0/2 - s_0$, if the invariant $u + 2s = 2s_0$ holds along the positive characteristics linking the two regions. These two families of negative characteristics cannot now be joined by an expansion fan and will intersect with each other, giving rise to a non-unique solution. If we insist that our model has a unique solution, then a break-down in the continuity of u and s must occur. That is, we allow a line of discontinuity in the x,t plane across which u and s are both discontinuous. To obtain a unique line of discontinuity we must pose relationships between the values of u and s on either side of the discontinuity. These relationships come from the physical conservation laws for the problem; their derivation will be discussed in the next section. Let us suppose that we are given two such relationships involving the values of u and s on either side of the discontinuity and that the speed of the discontinuity is U. Then the flow will consist of two uniform regions separated by a discontinuity at $x = -Ut$, and we have sufficient information to determine U and the value of s behind the discontinuity from the given relationships. This is shown in Figure 3.3. In general $u + 2s$ will not remain constant across the discontinuity which connects two uniform regions, one at rest and the other moving with speed U_0. We shall see in the next section that the height of the moving region is greater than that in the region at rest. Such a discontinuity is called a *hydraulic jump*.

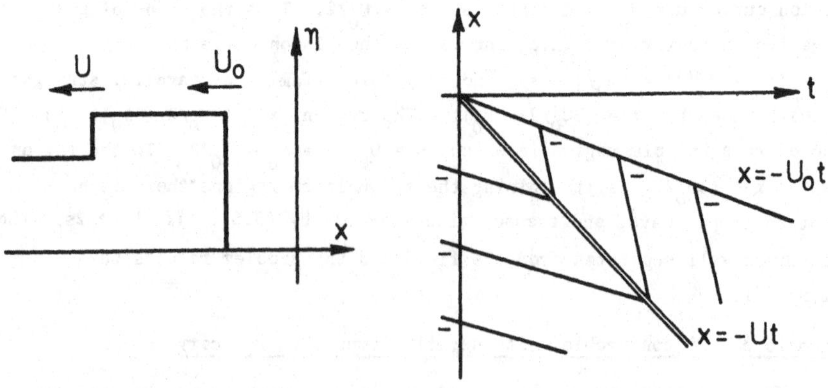

Figure 3.3. Impulsive piston problem; compressive case.

Example 4. Flow between reservoirs.

This is a variation on the dam break problem in which fluid at rest in $x < 0$ with initial height s_0^2/g is separated from a similar fluid at rest in $x > 0$ but with height s_1^2/g. At time $t = 0$ the barrier separating the fluids is removed, and the deeper fluid flows out into the shallower fluid. (They both have the same bottom $y = 0$.) There are now two regions at rest, one with characteristics of slope $\pm s_0$, the other with characteristics of slope $\pm s_1$, assumed to be less than s_0. These uniform regions will be bounded by $x = -s_0t$ and $x = s_1t$; the characteristic diagram is shown in Figure 3.4. A discontinuity must occur since, if it did not, in the intermediary region between the two uniform regions both $u + 2s = 2s_0$ and $u - 2s = -2s_1$. This would imply that both u and s are constants and that both families of characteristics are straight lines, which is not possible. Using a similarity argument we find that the discontinuity must be along a straight line, and the only possible configuration consists of three uniform regions and a centred simple wave.

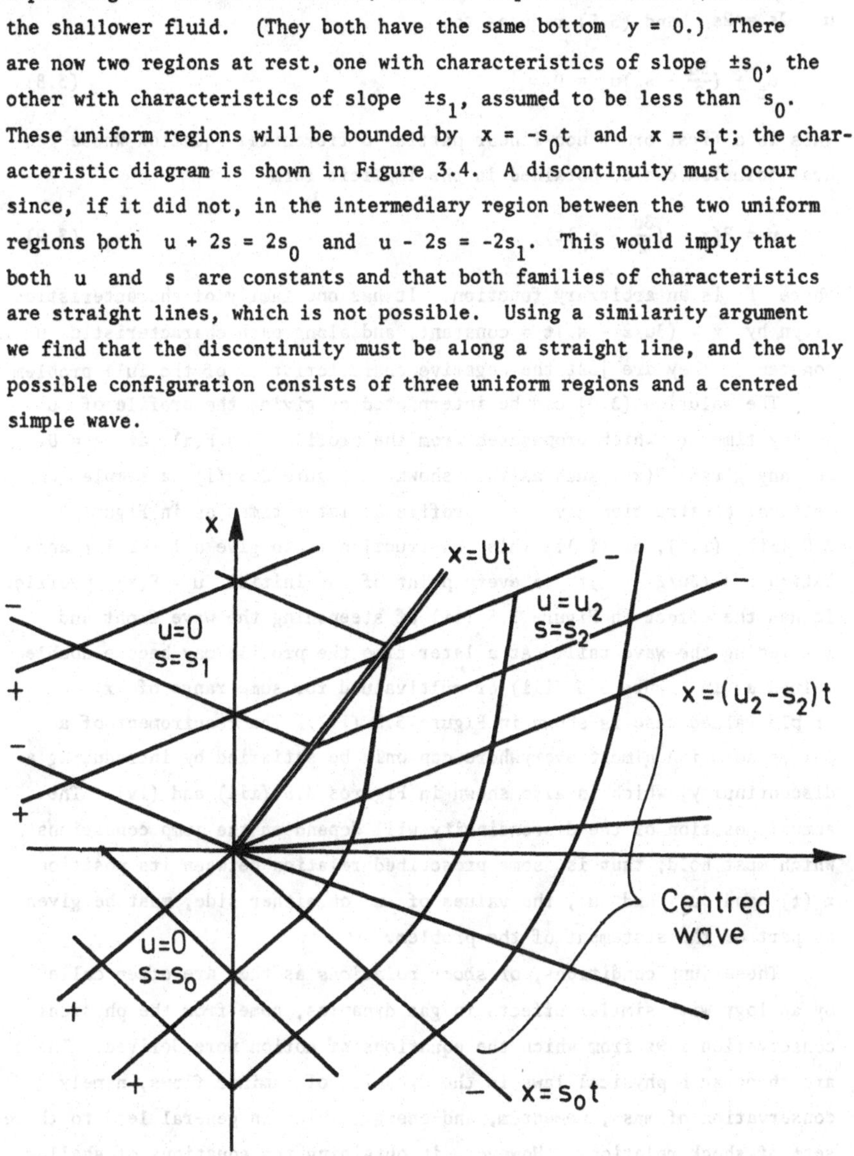

Figure 3.4. Flow between reservoirs.

2. HYDRAULIC JUMPS

A discussion of the onset of discontinuities in the solution of the shallow water theory model in a more general situation can be given when one of the integrals along the characteristics (i.e., one of the Riemann invariants) is constant everywhere. Thus we look for solutions in which $u + 2s = 2s_0$ and (3.5) reduces to

$$u_t + (\frac{3u}{2} - s_0)u_x = 0. \tag{3.8}$$

This is a first order non-linear partial differential equation whose general solution can be obtained in the implicit form

$$u = F(x - (\frac{3u}{2} - s_0)t), \tag{3.9}$$

where F is an arbitrary function. It has one family of characteristics given by $x - (3u/2 - s_0)t = \text{constant}$, and along each characteristic u is constant. They are just the negative characteristics of the full problem.

The solution (3.9) can be interpreted as giving the profile of u at any time t which propagated from the profile $u = F(x)$ at $t = 0$. For any given $F(x)$, such as that shown in Figure 3.5 (i), a simple geometrical construction gives the profile at later times as in Figures 3.5 (ii), (iii), or (iv). This construction is to give a lateral translation of $(3u/2 - s_0)t$ to every point of the initial $u = F(x)$ profile. It has the effect in Figure 3.5 (ii) of steepening the wave front and flattening the wave tail. At a later time the profile may become double valued as in Figure 3.5 (iii) or multivalued for some range of x. A triple valued case is shown in Figure 3.5 (iv). The requirement of a unique solution almost everywhere can only be satisfied by introducing a discontinuity, which is also shown in Figures 3.5 (iii) and (iv). The actual position of the discontinuity will depend on the jump conditions which must hold; that is, some prescribed relation between its position $x_s(t)$ and u_+ and u_-, the values of u on either side, must be given as part of the statement of the problem.

These jump conditions, or *shock relations* as they are often called by analogy with similar effects in gas dynamics, come from the physical conservation laws from which the equations of motion were derived. There are three such physical laws in the dynamics of laminar flows, namely conservation of mass, momentum, and energy, which in general lead to three sets of shock relations. However, in obtaining the equations of shallow water theory certain approximations have been made which have reduced the

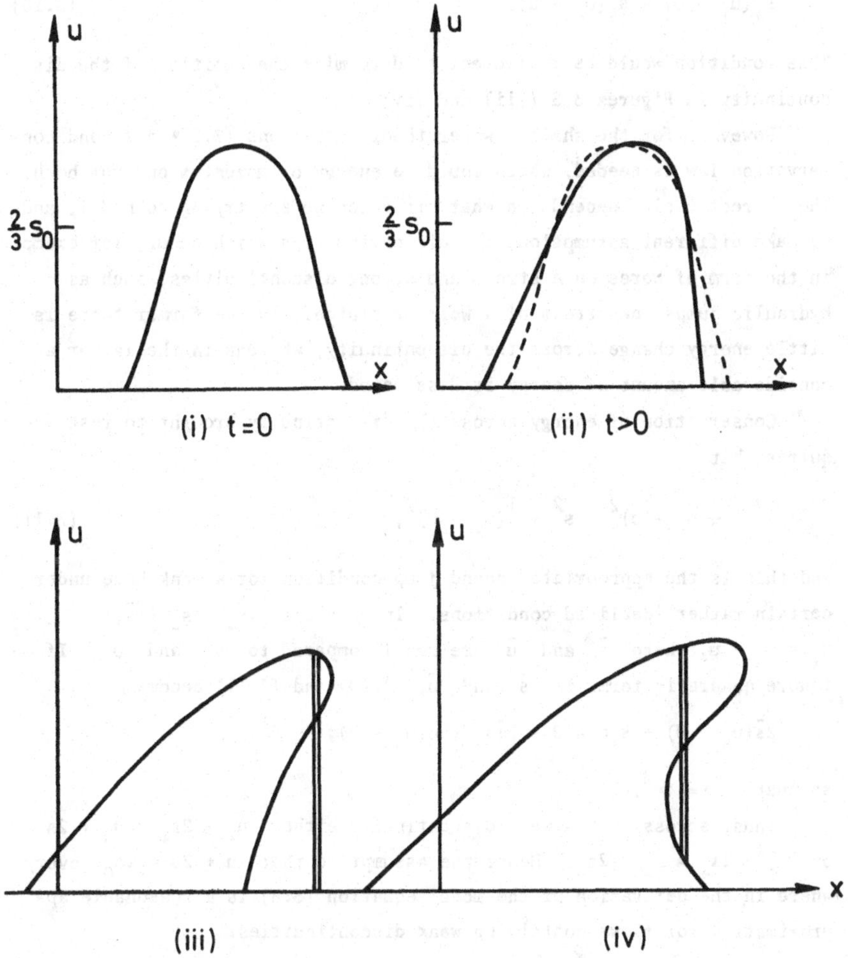

Figure 3.5. Nonlinear wave propagation.

order of the governing system of partial differential equations to two. Hence only two of the three conservation laws can be satisfied across the discontinuity. In the model equation (3.8) the order has been reduced further by the assumption that $u + 2s = 2s_0$, and only one conservation law can be applied. Conservation of mass is the obvious choice.

If we take moving axes, which reduces the discontinuity to rest, then velocities on either side of the discontinuity are $u_+ - U$ and $u_- - U$, where $U = dx_s/dt$. For the mass flow to be continuous,

$$s_+^2(u_+ - U) = s_-^2(u_- - U). \tag{3.10}$$

This condition would be sufficient to determine the position of the discontinuity in Figures 3.5 (iii) and (iv).

However, for the shallow water theory equations (2.27) a second conservation law is needed, which could be energy or momentum but not both. The correct choice depends on what situation we are trying to model, and we make different assumptions for discontinuities which occur, for example, in the form of bores on a stream and strong discontinuities, such as hydraulic jumps downstream of a weir or sluice. In the former there is little energy change across the discontinuity, whereas in the latter a considerable amount of energy is dissipated.

Conservation of energy across the discontinuity brought to rest requires that

$$s_+^2 + \frac{1}{2}(u_+ - U)^2 = s_-^2 + \frac{1}{2}(u_- - U)^2, \tag{3.11}$$

and this is the appropriate second jump condition for a weak bore under certain rather idealised conditions. In this case $s_+ = s_- + \bar{s}$, $u_+ = u_- + \bar{u}$, where $\bar{s}$ and $\bar{u}$ are small compared to s_- and u_-. If we ignore quadratic terms in $\bar{s}$ and $\bar{u}$, (3.10) and (3.11) become

$$2\bar{s}(u_- - U) + s_-\bar{u} = 0 - 2\bar{s}s_- + \bar{u}(u_- - U),$$

so that $\bar{u}^2 = 4\bar{s}^2$.

Thus, across such a weak discontinuity either $u_+ + 2s_+ = u_- + 2s_-$ or $u_+ - 2s_+ = u_- - 2s_-$. Hence the assumption that $u + 2s = 2s_0$ everywhere in the derivation of the model equation (3.8) is a reasonable approximation for flows containing weak discontinuities.

Conservation of momentum is, however, more appropriate for most of the flows discussed using the shallow water theory. The derivation of the jump condition is not quite as obvious as for conservation of mass and energy. More sophisticated methods of constructing the correct conservation laws will be discussed in the context of gas dynamics in Chapter V. A simple argument is that relative to the discontinuity the quantity conserved is

$$(\text{mass flow})(\text{velocity}) + \int_0^h (p-p_0)dy,$$

where $p - p_0 = \rho g(h-y)$. This reduces to

$$s_+^2(u_+ - U)^2 + \frac{1}{2}s_+^4 = s_-^2(u_- - U)^2 + \frac{1}{2}s_-^4. \tag{3.12}$$

We can now complete the compressive piston problem in the previous section on the assumption that a hydraulic jump forms. The conditions in front of the discontinuity are $u = 0$, $s = s_0$ and behind are $u = U_0$ with $s = s_1$ unknown. Using (3.10) and (3.12), the relations

$$s_0^2 U = s_1^2 (U - U_0),$$
$$s_0^2 U^2 + \frac{1}{2} s_0^4 = s_1^2 (U - U_0)^2 + \frac{1}{2} s_1^4 \tag{3.13}$$

determine the speed U of the hydraulic jump and s_1^2/g, the height behind it, in terms of the piston speed U_0 and initial height s_0^2/g. There will be a rate of energy loss $[s_0^2 - s_1^2 + \frac{1}{2}U^2 - \frac{1}{2}(U-U_0)^2]s_0^2 U$, which after some manipulation can be simplified to the form

$$\frac{U}{4s_1^2}(s_1^2 - s_0^2)^3. \tag{3.14}$$

Thus, energy will only be dissipated by the hydraulic jump with $U > 0$ if s_1 is greater than s_0. By superimposing a constant velocity $-U$ on this flow we see that a stationary hydraulic jump (and a weak bore) can only change a supercritical stream $(U > s_0)$ to a subcritical one $(U-U_0 < s_1)$ and not vice-versa.

The shallow water theory model can be extended to motion over a bottom of variable slope, and this is discussed in the next section. The profile of a disturbance propagating onto a beach is found to steepen in the qualitative way described in this section, and eventually the wave breaks. The actual motion is then turbulent and is not sensibly modelled by a hydraulic jump, but it is this breakdown of the continuous wave profile that enables the energy of the incoming waves to be dissipated.

3. SOLITARY AND CNOIDAL WAVES

We have examined in Chapter II.3 and II.4 the gravity wave problem in two special situations when one or other of the parameters ε and δ defined in Chapter II.2 was very small. We now examine this limiting procedure more carefully and discover that when both ε and δ are small the motion depends on the relative size of ε and δ^2. The full gravity wave model is given by (2.3), (2.4), (2.5) and (2.6), and we consider only two dimensional flow over a fixed horizontal bottom $y = -h_0$. The initial profile at $t = 0$ we redefine as $\eta = \eta_0 F(x/\lambda)$, so that λ and η_0 are the length scales of the disturbance in the horizontal and vertical directions. There are then four dimensional parameters in the problem, namely g, h_0, η_0 and λ, and from them we can construct two independent non-dimensional parameters: $\varepsilon = \eta_0/h_0$ and $\delta = h_0/\lambda$. The problem can be made non-dimensional by defining $x = \lambda\bar{x}$, $y = h_0\bar{y}$, $\eta = \eta_0\bar{\eta}$, $t = t_0\bar{t}$, $\phi = \phi_0\bar{\phi}$ where t_0 and ϕ_0 are to be determined from g, h_0, η_0, λ or equivalently g, h_0, ε, δ. Thus we write

$$t_0 = \omega(\varepsilon,\delta)\left(\frac{h_0}{g}\right)^{\frac{1}{2}}, \qquad \phi_0 = \kappa(\varepsilon,\delta)\sqrt{gh_0^3}, \tag{3.15}$$

where ω and κ are unknown functions of ε and δ.

The problem then becomes (and without ambiguity the bars on the non-dimensional variables can be omitted)

$$\phi_{yy} + \delta^2\phi_{xx} = 0; \tag{3.16}$$

$$\text{on} \quad y = -1, \quad \phi_y = 0; \tag{3.17}$$

$$\text{on} \quad y = \varepsilon\eta, \quad \eta + \frac{\kappa}{\varepsilon\omega}\phi_t + \frac{1}{2}\frac{\kappa^2}{\varepsilon}(\phi_y^2 + \delta^2\phi_x^2) = 0, \tag{3.18a}$$

$$\eta_t - \frac{\omega\kappa}{\varepsilon}\phi_y + \delta^2\omega\kappa\phi_x\eta_x = 0; \tag{3.18b}$$

and

$$\text{on} \quad t = 0, \quad \phi = 0 = \phi_t, \quad \eta = F(x). \tag{3.19}$$

We now examine simplifications of this system for limiting values of ε and δ with appropriate values of κ and ω.

(i) $\underline{\varepsilon \ll 1,\ \delta = 1}$: Small Amplitude Waves. To obtain non-trivial free surface conditions (3.18) we must choose $\kappa/\varepsilon\omega = 1$ and $\omega\kappa/\varepsilon = 1$ so that $\kappa = \varepsilon$, $\omega = 1$. The first term in our expansion in powers of ε leads to the Stokes wave problem since (3.18) reduces to (2.7). The solution will be valid for time scales of order t_0, that is, time scales $O((h_0/g)^{\frac{1}{2}})$, and the velocity potential will be $O(\varepsilon\sqrt{gh_0^3})$. For large enough times the Stokes wave solution will not be a valid approximation. Since, however, on small disturbance theory a general disturbance decays like $t^{-\frac{1}{2}}$, this lack of uniform validity is not very interesting.

(ii) $\underline{\delta \ll 1,\ \varepsilon = 1}$. Shallow Water Approximation. From (3.16) and (3.17), $\phi_y \sim O(\delta^2)$ although $\phi \sim O(1)$. Correct to $O(\delta^2)$ these equations lead to

$$\phi = a(x,t) + \delta^2[b(x,t) - (y+\tfrac{1}{2}y^2)a_{xx}]. \tag{3.20}$$

To obtain a non-trivial problem we must choose ω and κ so that (3.18b) does not reduce to $\eta_t = 0$. Hence choose $\omega\kappa\delta^2 = 1$ so that all the terms in (3.18b) are $O(1)$. From (3.18a) choose $\kappa = \omega$ so that $\kappa = \omega = 1/\delta$, and the time scale of this problem is $O((\lambda^2/gh_0)^{\frac{1}{2}})$. Substituting (3.20) into (3.18) and expanding in powers of δ, we find that the first term in the expansion satisfies the shallow water equations as in (3.3) and (3.4). (Note that the origin of y is different in the two derivations). Because of the time scale of their validity they may not represent the initial stages of a flow accurately.

If the bottom has a depth variation with a horizontal length scale the same as that of the surface disturbance, then (3.17) becomes

$$\text{on } y = -h(x), \quad \phi_y = -\delta^2\phi_x h'$$

and (3.20) becomes

$$\phi = a(x,t) + \delta^2[b(x,t) - \{y(ha_x)_x + \frac{y^2}{2} a_{xx}\}].$$

Substituting into (3.18) and taking the first term in an expansion in powers of δ, we obtain

$$u_t + \eta_x + uu_x = 0,$$

$$\eta_t + (uh)_x + \eta u_x + u\eta_x = 0.$$

A simpler form is obtained by writing $\eta + h = H$, so that

$$u_t + H_x + uu_x = h', \qquad H_t + uH_x + Hu_x = 0. \tag{3.21}$$

With $h = 1$, these reduce to the shallow water equations (3.3) and (3.4) in dimensional coordinates. The effect of the variable bottom is to introduce a nonhomogeneous term into the shallow water equations, which considerably complicates their solution and gives rise to solutions of a very different form.

(iii) <u>$\delta \ll 1$, $\varepsilon \ll 1$.</u> <u>Small amplitude shallow water waves.</u> If we choose $\kappa = \varepsilon/\delta$ and $\omega = 1/\delta$, then ϕ has the form (3.20), and on substituting into the free surface conditions, we find that

$$u_t + \eta_x = 0 = \eta_t + u_x \tag{3.22}$$

is the limiting form as ε and $\delta \to 0$. These are the equations of tidal waves and are valid for times $O((\lambda^2/gh_0)^{\frac{1}{2}})$. They are not dispersive, and any given profile propagates without change of shape with dimensional speed

$\sqrt{gh_0}$. They are therefore limiting forms of both shallow water and Stokes waves. For times greater than this, the equations (3.22) will be modified and derivatives with respect to x will become larger than those with respect to t. Hence we have to include higher order terms in the expansion for ϕ, which has the form, correct to $O(\delta^4)$,

$$\phi = a(x,t) + \delta^2[b(x,t) - (y+\tfrac{1}{2}y^2)a_{xx}] + \delta^4\Big[c(x,t) - (y+\tfrac{1}{2}y^2)b_{xx} + \Big(\frac{y^4}{24} + \frac{y^3}{6} - \frac{y}{3}\Big)a_{xxxx}\Big].$$

The free surface conditions now become

$$\eta_x + u_t + \varepsilon u u_x = O(\varepsilon\delta^2),$$

$$\eta_t + u_x + \varepsilon u\eta_x + \varepsilon\eta u_x + \frac{\delta^2}{3}u_{xxx} = O(\delta^4) \quad \text{or} \quad O(\varepsilon\delta^2).$$

Clearly we can distinguish three cases: $\varepsilon >> \delta^2$, $\varepsilon << \delta^2$ and $\varepsilon \sim \delta^2$; the latter is the most interesting, and we therefore consider $\varepsilon = \delta^2$.

Consider a tidal wave propagating with non-dimensional speed +1. After a long time t the scale of $x-t$ will still be $O(1)$ since the profile does not change. Hence the correct scaling for a long time situation is obtained by writing $z = x-t$ and $\tau = t\varepsilon^n$, where $n > 0$. The equations then become

$$\eta_z - u_z + \varepsilon^n u_\tau + \varepsilon u u_z = O(\varepsilon^2),$$

$$u_z - \eta_z + \varepsilon^n\eta_\tau + \varepsilon(u\eta_z + \eta u_z + \frac{1}{3}u_{zzz}) = O(\varepsilon^2).$$

To obtain a non-trivial solution choose $n = 1$ and eliminate u by writing $u = \eta + O(\varepsilon)$. The arbitrary function of τ is zero if we are only considering a wave moving to the right. Then

$$\eta_\tau + \frac{3}{2}\eta\eta_z + \frac{1}{6}\eta_{zzz} = 0. \tag{3.23}$$

This is called the Korteweg-DeVries equation and is valid for times $O((h_0/g)^{\frac{1}{2}}/\delta^3)$ when $\varepsilon = \delta^2 << 1$. It represents a balance between the linear tidal wave term η_τ, the non-linear shallow water term $\eta\eta_z$, and the dispersive Stokes wave term η_{zzz}. If $\delta^2 << \varepsilon$ the dispersive term is not present and we have a non-linear simple wave as discussed in Chapter III.1, with the solution given by equation (3.9). If $\varepsilon << \delta^2$ we have a linear dispersive system with dimensional wave speed $(1 - k^2h_0^2/6)\sqrt{gh_0}$. It is interesting to note that this is the same as the first two terms in the expansion of the Stokes wave dispersion relation (2.11) for small kh_0.

The Korteweg-DeVries equation has travelling wave type solutions which can be obtained by writing $\eta = f(z-C\tau)$. Substituting this in (3.23) and integrating once gives

$$\frac{d^2f}{ds^2} = 6Cf - \frac{9}{2}f^2 + A, \tag{3.24}$$

where A and C are constants and $s = z - C\tau = x - (1+C\varepsilon)t$. One very special solution of (3.24) is obtained by requiring that as $s \to \pm\infty$, f and its derivatives tend to zero. Then $A = 0$ and

$$\left(\frac{df}{ds}\right)^2 = 6Cf^2 - 3f^3. \tag{3.25}$$

It is easily verified that a solution is

$$f = \alpha \operatorname{sech}^2 \beta s, \tag{3.26}$$

where $\alpha = 2C$ and $\beta = \sqrt{3C/2}$. This profile is called a *solitary wave*, and its speed and shape depend solely on the one parameter C. A general profile in a shallow channel will eventually break up into propagating solitary waves. Each solitary wave is very stable and can be observed both in a canal and in the laboratory.

This is not the only solution; a general integral of (3.25) involves cnoidal functions cn(s). To examine their properties we perform a phase plane treatment of (3.24) with $A = 0$ and $g = df/ds$. Then

$$\frac{dg}{df} = \frac{6Cf - 9/2f^2}{g}. \tag{3.27}$$

The critical points are $f = 0$, $g = 0$ and $f = 4C/3$, $g = 0$; their character is as shown in Figure 3.6 for $C > 0$. Each closed curve in the (f,g) diagram represents a possible integral path with s as the parameter defining position along it. The solitary wave solution is the closed curve through the origin which returns to the origin as s ranges from $-\infty$ to ∞. A closed curve internal to this singular one will be traversed in a finite range of s and hence represents a solution periodic in s called a *cnoidal wave*. For a given C there is a whole family of possible cnoidal waves. A curve external to the singular one will have unbounded f for some value of s and cannot represent a wave. With A non-zero in (3.24) the phase plane will be similar provided that $2C^2 + A > 0$, but there will be no bounded solutions if $2C^2 + A \leq 0$.

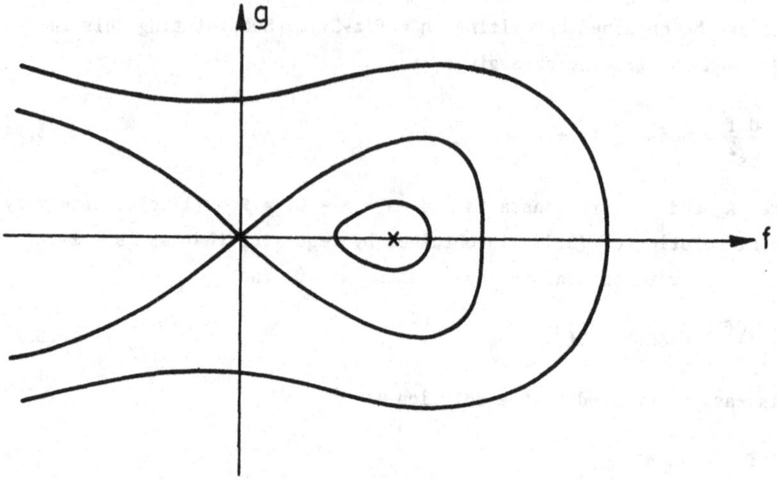

Figure 3.6. Phase plane for solitary and cnoidal waves.

EXERCISES

1. Incompressible fluid of depth s_0^2/g is contained in $-\infty < x < 0$ and is separated by a sluice from fluid of depth s_1^2/g and the same horizontal bottom $y = 0$ in $0 < x < \infty$. At time $t = 0$ the sluice is instantaneously removed and the first reservoir discharges into the second $(s_0 > s_1)$. Draw the characteristic diagram in the x,t plane and hence show that a discontinuity is propagated into the second reservoir. If the height of the fluid behind this discontinuity is s_2^2/g and $s_2 < 2/3\ s_0$, show that at $x = 0$ the depth is always $4s_0^2/9g$ and the discharge rate is $8s_0^3/(27g)$. If $s_2 > 2/3\ s_0$ show that the discharge rate decreases from this value as s_2 increases.

2. A surface wave propagates along a straight horizontal channel into water initially at rest with height h_0. At $t = 0$ the wave has the form

$$h(x,0) = \begin{cases} h_0(1 + 2\alpha) & x \le 0 \\ h_0(1 + \alpha + \alpha \cos \pi x/a) & 0 \le x \le a \\ h_0 & x \ge a, \end{cases}$$

where α is small and positive. Show that the wave will break at time $2a/(3\pi\alpha\sqrt{gh_0})$, approximately.

3. In the presence of a fixed sluice-gate, the depth of steady open channel flow changes from h_1 upstream to γh_1 downstream, where $\gamma < 1$. The upstream velocity is u_1. Assuming that no energy is lost at the transition, show that

$$\gamma = \tfrac{1}{4}[F_1^2 + F_1(8 + F_1^2)^{\frac{1}{2}}],$$

where $F_1 = u_1/\sqrt{gh_1} < 1$ is the upstream Froude number. Obtain an expression in terms of γ for the horizontal force needed to hold the sluice-gate in place.

4. Show that the shallow water equations for two-dimensional flow over a bottom of slope m can be written in the form

$$u \pm 2s - mgt = \text{constant on } \frac{dx}{dt} = u \pm s,$$

where $s^2 = g(\eta + mx)$. For a generalized simple wave flow the Riemann invariant $u+2s - mgt$ is constant everywhere. Show that the solution is then

$$u - mgt = F(x - \frac{3}{2}ut + s_0t + mgt^2),$$

where

$$u = F(x) \quad \text{and} \quad s = s_0 - \tfrac{1}{2}F(x) \quad \text{at} \quad t = 0.$$

5. For the flow in question 4, $F(x)$ is a continuous positive function which decays to zero as $|x| \to \infty$. Show that the solution is continuous for $t < t_c$, where t_c satisfies

$$2 + 3t_cF'(\alpha_c) = 0 = F''(\alpha_c).$$

If $F(x) = 2x - x^2$ in $0 < x < 2$ and is zero otherwise, show that $t_c = 1/3$.

6. A fluid of depth h is in two-dimensional rotational motion with $u = u_0(y)$, $v = 0$ when it is subjected to a disturbance of wavelength λ. Assuming that $\lambda >> h$, show that the flow satisfies the equations

$$\frac{\partial u}{\partial x} + \frac{\partial v}{\partial y} = 0$$

$$\frac{\partial u}{\partial t} + u\frac{\partial u}{\partial x} + v\frac{\partial u}{\partial y} = -g\frac{\partial \eta}{\partial x}$$

with $v = 0$ on $y = 0$ and $\partial\eta/\partial t + u\,\partial\eta/\partial x = v$ on $y = \eta(x,t)$. Show that a small amplitude travelling wave of the form $\eta = h(1 + \varepsilon e^{ik(x-ct)})$ is possible if

$$\int_0^h \frac{g\,dy}{(c - u_0(y))^2} = 1.$$

Find the possible values of c when $u_0 = ky$ and comment on the fact that $c = u_0$ for two values of y in $(0,h)$.

7. a) Show that the shallow water equations for three dimensional flow (with ζ as the other horizontal axis and w the corresponding velocity) are:

$$u_t + uu_x + wu_x + 2ss_x = 0,$$

$$w_t + uw_x + ww_\zeta + 2ss_\zeta = 0,$$

$$2s_t + 2us_x + 2ws_\zeta + su_x + sw_\zeta = 0, \quad \text{where} \quad s^2 = g\eta.$$

b) Follow the derivation for the Korteweg-DeVries equation allowing variations in the other horizontal direction ζ. Assuming that length scales for ζ and x are comparable but that the wave is propagating in the x direction, show that the equation is

$$\frac{\partial}{\partial z}\left(\eta_\tau + \frac{3}{2}\eta\eta_z + \frac{1}{6}\eta_{zzz}\right) + \eta_{\zeta\zeta} = 0$$

where $z = x-t$.

Chapter IV
Compressible Flow

1. INTRODUCTION

In many gas flows which do not involve extreme pressures or temperatures, compressibility is an important effect while viscosity and other real gas effects may be neglected away from any rigid boundaries. Almost all the theory contained in the following three chapters will be concerned with an inviscid ideal gas as defined in Chapter I. The only exception which we shall consider occurs in the study of shock waves, where viscosity is important in the interior of the shock. Some reference will also be made to other real gas effects, such as radiation and dissociation, which can be important under certain extreme conditions.

The equations for compressible flow of an ideal gas were developed in Section I.4. Most of the flows considered here will be homentropic, hence P/ρ^{γ} will be constant throughout the fluid; this is an example of a *barotropic* flow. We shall also neglect the effects of external body forces.

Sound Waves

The difference between incompressible and compressible flows is typified by consideration of the way in which a 'sound wave' travels in the fluid. A sound wave is a disturbance in which the changes in the flow variables are of small amplitude compared to their ambient values. Suppose that an inviscid compressible gas is at rest at constant pressure p_0 and density ρ_0 and that a small disturbance causes the pressure to become $p_0 + p_1$, the density $\rho_0 + \rho_1$ and the velocity q_1, where $p_1 \ll p_0$, $\rho_1 \ll \rho_0$, and $q_1 \ll \sqrt{p_0/\rho_0}$, which is the only quantity with the dimensions of velocity in the undisturbed state. The equation of continuity

(1.6) may be written as

$$\frac{\partial\rho_1}{\partial t} + (\rho_0+\rho_1)\ \mathrm{div}\ q_1 + (\mathrm{grad}\ \rho_1)\cdot q_1 = 0$$

or, neglecting products of small quantities,

$$\frac{\partial\rho_1}{\partial t} + \rho_0\ \mathrm{div}\ q_1 = 0. \tag{4.1}$$

Similarly, to the same approximation, Euler's equations (1.8) become

$$\frac{\partial q_1}{\partial t} + \frac{1}{\rho_0}\ \mathrm{grad}\ p_1 = 0, \tag{4.2}$$

and the energy equation (1.35) leads to

$$\frac{\partial S}{\partial t} = 0, \quad \text{if} \quad \nabla S = 0 \quad \text{at} \quad t = 0.$$

Thus, S is a constant S_0 for all space and time in this approximation, and p may be expanded in terms of ρ_1 to give

$$p_1 = \left(\frac{\partial p}{\partial\rho}\right)_0 \rho_1, \tag{4.3}$$

where p is regarded as a function of ρ and S and $\left(\frac{\partial p}{\partial\rho}\right)_0$ is evaluated in the undisturbed state with $S = S_0$. From equations (4.1), (4.2) and (4.3)

$$\frac{\partial^2\rho_1}{\partial t^2} = \left(\frac{\partial p}{\partial\rho}\right)_0 \nabla^2\rho_1, \tag{4.4}$$

so that ρ_1, and similarly p_1 and q_1, satisfies the wave equation. The speed of sound with which the disturbance is propagated is a_0, where

$$a_0^2 = \left(\frac{\partial p}{\partial\rho}\right)_0.$$

For an ideal gas, from (1.36), the entropy S is a function of p/ρ^γ and $a_0^2 = \gamma(p_0/\rho_0)$. In general, the speed of sound, a, at any point of the flow field is defined by $a^2 = \frac{dp}{d\rho} = \frac{\gamma p}{\rho}$ for isentropic flow.

This result is in contrast to the situation in an incompressible fluid, where a small disturbance affects the whole flow field instantaneously and the velocity of sound is therefore infinite.

For a given flow there are two reference speeds of sound which will be useful in the later discussion. The *stagnation speed of sound,* a_0, is the speed of sound which would exist in the gas if it were brought to rest isentropically. Hence, Bernoulli's equation (1.37) for steady flow can be written in terms of a as

$$\frac{1}{2} q^2 + \frac{a^2}{\gamma-1} = \frac{a_0^2}{\gamma-1} . \tag{4.5}$$

If we define the *Mach number* of the flow by q/a, the *critical speed of sound* a_* is the speed of sound which would exist in the gas if it were brought to Mach number one isentropically. Thus, Bernoulli's equation can alternatively be written

$$\frac{1}{2} q^2 + \frac{a^2}{\gamma-1} = \frac{\gamma+1}{2(\gamma-1)} a_*^2. \tag{4.6}$$

The significance of the Mach number can be seen from (4.5) since

$$\frac{(\gamma-1)}{2} M^2 + 1 = \frac{a_0^2}{a^2} = \left(\frac{\rho_0}{\rho}\right)^{\gamma-1} = \left(\frac{p_0}{p}\right)^{(\gamma-1)/\gamma} = \frac{M^2 a_0^2}{q^2} \tag{4.7}$$

on using $p/p_0 = (\rho/\rho_0)^\gamma$ and $a^2 = \gamma p/\rho$. Thus, all the flow variables can be expressed in terms of their stagnation values and the Mach number.

The Mach Number

We now look for similarity properties of the equations of flow by transforming to non-dimensional variables and finding which non-dimensional parameters are important. Suppose that the flow is specified by a typical velocity U, density ρ_1, length L and speed of sound a_1. We introduce non-dimensional variables by writing

$$q = U\hat{q}, \quad x = L\hat{x}, \quad \rho = \rho_1\hat{\rho}, \quad a = a_1\hat{a}, \quad t = \frac{L\hat{t}}{U},$$

and, since $a^2 = \gamma p/\rho$, $p = \rho_1 a_1^2 \hat{p}$. The equations (1.6) and (1.8) become

$$\frac{\partial\hat{\rho}}{\partial\hat{t}} + \operatorname{div}(\hat{\rho}\hat{q}) = 0$$

and

$$\frac{\partial\hat{q}}{\partial\hat{t}} + (\hat{q}\cdot\nabla)\hat{q} + \frac{a_1^2}{U^2} \operatorname{grad} \hat{p} = 0.$$

In homentropic flow $\hat{p}/\hat{\rho}^\gamma$ is constant, and we see that the flow depends only on the boundary conditions and on the parameter U/a_1, which is the Mach number in the reference state. Thus, the flow of a stream past two similar bodies will be similar if the free stream Mach number is the same in both cases.

We next consider a point source of sound, such as a small obstacle moving with constant velocity U through a uniform gas at rest. Suppose that the speed of sound in the undisturbed gas is a_1. Then there is a basic difference in the resulting sound field depending on whether the Mach number $M_1 = U/a_1$ is greater or less than one. If $M_1 < 1$, the source is moving *subsonically* and the effect of the disturbance will travel through the gas faster than the source can travel, so that eventually the whole flow field will be influenced by the presence of the source. However, if $M_1 > 1$, the source moves *supersonically* and the disturbance due to the source will travel more slowly than the source itself. Thus, the presence of the source will influence the flow only within a circular cone which trails behind the source. This cone is called the *Mach cone*, and its semi-angle, which is given by $\mu_1 = \sin^{-1}(1/M_1)$, is the *Mach angle*.

In a similar way, a stream with speed U is said to be subsonic or supersonic as $U < a$ or $U > a$. If a velocity $-U$ is superimposed on the whole system, the source is brought to rest but the pattern of the flow is unchanged. Thus, if a small obstacle is placed in a subsonic stream the whole fluid will be affected by its presence. On the other hand, if the stream is supersonic the flow will be affected only within the Mach cone whose vertex is at the obstacle. This cone is called the *region of influence* of the obstacle, and all other points of the flow are in the *zone of silence*. In two dimensions there is a corresponding difference between subsonic and supersonic flow. For supersonic flow the effect of a small obstacle is felt only *on the Mach lines*.† The region of influence of a small obstacle in two and three dimensions is illustrated in Figure 4.1.

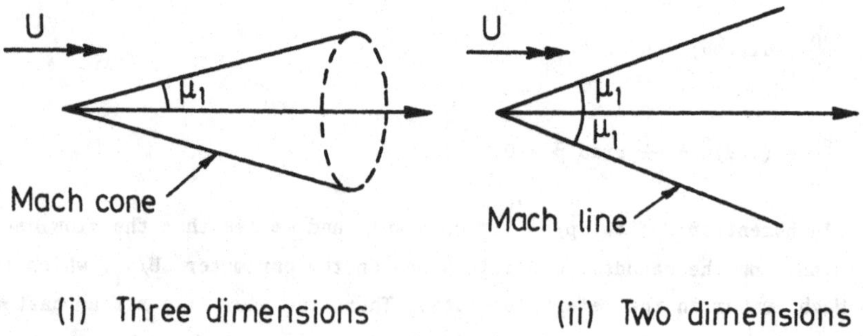

Figure 4.1. Supersonic flow past a small obstacle.

---

†The difference between the two cases is discussed in more detail in Section VI.2.

The difference between subsonic and supersonic flow may be identified with the fact that the equations of motion are hyperbolic in the supersonic case and elliptic in the subsonic case. This will be discussed in more detail later.

2. NOZZLE FLOW

If we consider a long straight channel or pipe whose cross-section varies only slowly along its length, we can model the steady flow down the pipe by assuming that it is virtually one-dimensional. This is equivalent to neglecting the velocity components perpendicular to the direction of mean flow (which is taken in the x direction) and averaging the velocity component u over a cross-section of the pipe. Then, if $A(x)$ is the cross-sectional area, continuity implies that

$$\rho A u = Q, \tag{4.8}$$

where Q is the constant mass flux. Using (4.7) we can write ρ and u in terms of the Mach number M so that

$$A = \frac{Q}{\rho_0 a_0}\frac{1}{M}\left(1 + \frac{\gamma-1}{2}M^2\right)^{(\gamma+1)/2(\gamma-1)} \tag{4.9}$$

Thus

$$\frac{1}{A}\frac{dA}{dx} = \frac{M^2 - 1}{M(1 + \frac{\gamma-1}{2}M^2)}\frac{dM}{dx},$$

and if A has a minimum it can only occur if either $M = 1$ or $dM/dx = 0$. Plotting A against M, as in Figure 4.2, we see that A has a minimum value

$$A_c = \frac{Q}{a_0\rho_0}\left(\frac{\gamma+1}{2}\right)^{(\gamma+1)/2(\gamma-1)} \tag{4.10}$$

and continuous flows are only possible if $A \geq A_c$ throughout the pipe. A practical situation is a convergent-divergent pipe called a Laval nozzle. Such a nozzle is attached to a large reservoir of gas at rest and the gas is allowed to flow out through the nozzle. The flow is determined by the pressure which is applied at the downstream end of the nozzle. From (4.7)

$$p/p_0 = \left(1 + \frac{\gamma-1}{2}M^2\right)^{-\gamma/(\gamma-1)}$$

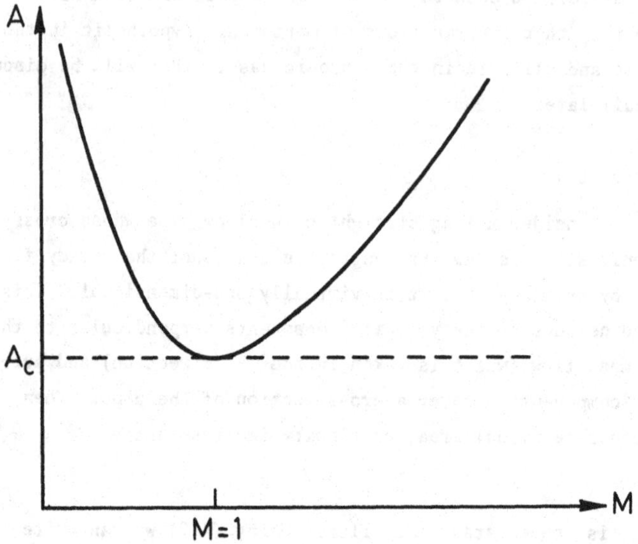

Figure 4.2. Variation of cross-sectional area with Mach number.

and the exit pressure determines the Mach number at exit, which from (4.9) gives the value of Q.

For a given Q, there are two possible types of continuous flow depending on whether the minimum value, A_m, of A is greater or equal to A_c. If $A_m > A_c$ then it can be seen from Figure 4.2 that as A decreases to A_m, M increases from zero to a maximum value which is less than one. In the divergent part of the nozzle, M decreases and the flow is always subsonic, as shown in Figure 4.3 (i). However, if $A_m = A_c$ the flow becomes sonic at the throat, where $A = A_m$, and it is possible for the flow in the divergent part of the nozzle to be supersonic, as in Fig. 4.3 (ii), if the pressure at the exit has the appropriate value. This transition from subsonic to supersonic flow is a phenomenon which is used in the design of supersonic wind tunnels. If $A_m < A_c$, then a continuous flow of this kind is not possible; the channel becomes "choked" and shock waves will form in the divergent part of the nozzle.

It is also possible to use a Laval nozzle to decelerate flow from supersonic to subsonic, and again sonic conditions will occur at the throat.

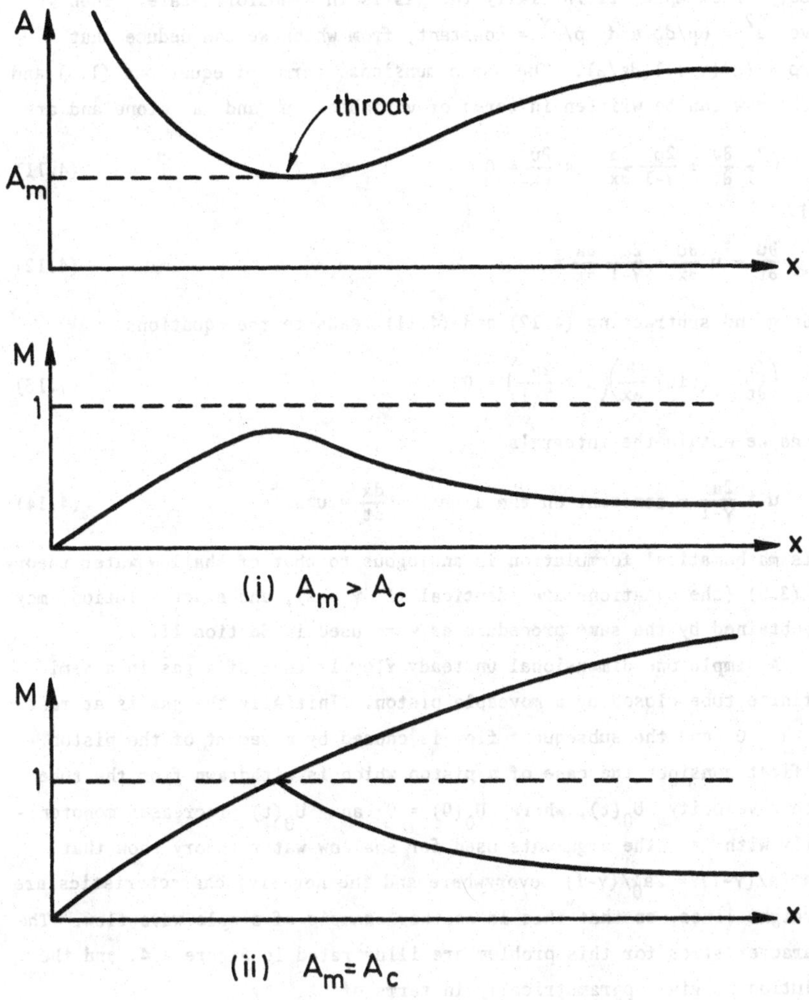

Figure 4.3. Flow conditions in a nozzle.

3. ONE DIMENSIONAL UNSTEADY FLOW

We now consider some exact solutions of the equations of motion of an inviscid fluid. The first example to be studied is one dimensional unsteady flow. We assume that the flow is homentropic, which will be the case, for example, if initially the gas is in a uniform state. Then we have $a^2 = dp/d\rho$ and p/ρ^γ = constant, from which we can deduce that $d\rho/\rho = (2/(\gamma-1))(da/a)$. The one dimensional forms of equations (1.6) and (1.8) now can be written in terms of variables u and a alone and are

$$\frac{2}{\gamma-1}\frac{\partial a}{\partial t} + \frac{2u}{\gamma-1}\frac{\partial a}{\partial x} + a\frac{\partial u}{\partial x} = 0 \tag{4.11}$$

and

$$\frac{\partial u}{\partial t} + u\frac{\partial u}{\partial x} + \frac{2a}{\gamma-1}\frac{\partial a}{\partial x} = 0. \tag{4.12}$$

Adding and subtracting (4.12) and (4.11) leads to the equations

$$\left(\frac{\partial}{\partial t} + (u\pm a)\frac{\partial}{\partial x}\right)\left(u \pm \frac{2a}{\gamma-1}\right) = 0. \tag{4.13}$$

Hence we obtain the integrals

$$u \pm \frac{2a}{\gamma-1} = \text{constant on the lines} \quad \frac{dx}{dt} = u\pm a.^\dagger \tag{4.14}$$

This mathematical formulation is analogous to that of shallow water theory in (3.5) (the equations are identical if $\gamma = 2$), and exact solutions may be obtained by the same procedure as were used in Section III.1.

A simple one dimensional unsteady flow is that of a gas in a semi-infinite tube closed by a moveable piston. Initially the gas is at rest in $x \leq 0$ and the subsequent flow is caused by movement of the piston. We first consider the case of a piston which is withdrawn from the tube with a velocity $U_0(t)$, where $U_0(0) = 0$ and $U_0(t)$ increases monotonically with t. The arguments used for shallow water theory show that $u + 2a/(\gamma-1) = 2a_0/(\gamma-1)$ everywhere and the negative characteristics are straight lines, so that this is another example of simple wave flow. The characteristics for this problem are illustrated in Figure 4.4, and the solution is given parametrically in terms of t_0 by

$$\begin{aligned} u &= U_0(t_0), \quad a = a_0 - \frac{\gamma-1}{2}U_0(t_0), \\ x &= \int_0^{t_0} U_0\,dt + (t-t_0)\left(\frac{\gamma+1}{2}U_0(t_0) - a_0\right), \end{aligned} \tag{4.15}$$

---

† An alternative method for obtaining these Riemann Invariants is given in the Appendix.

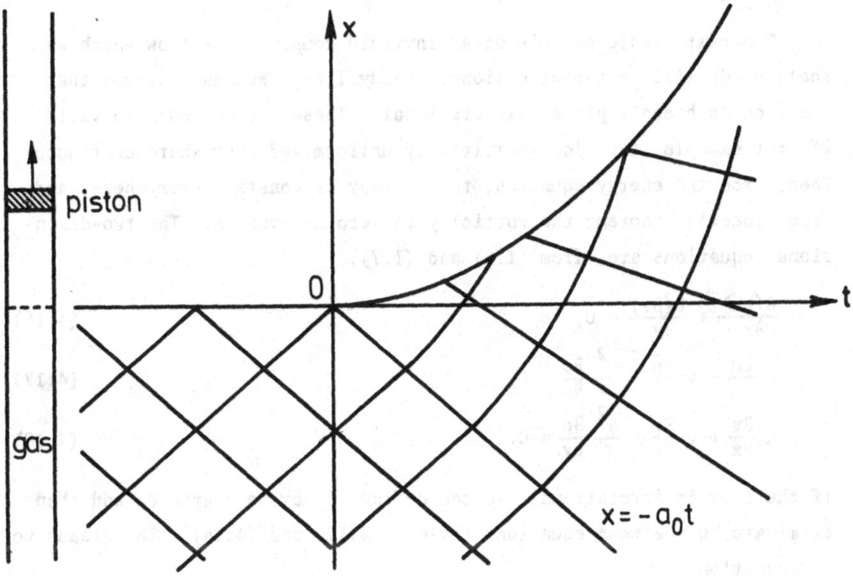

Figure 4.4. Expanding piston problem.

if $-a_0t \leq x \leq \int_0^t U_0(\tau)d\tau$. If $x < -a_0t$ the flow is undisturbed.

This solution is only possible if the speed of the piston is not too large. The speed of sound must be positive, and from (4.15) we must have

$$U_0(t) \leq \frac{2a_0}{\gamma-1}.$$

If $U_0(t)$ exceeds this value, a cavity will form behind the piston and the gas will expand freely in such a way that it is bounded by the characteristic with slope $2a_0/(\gamma-1)$ on which $a = 0$.

An analytic solution can also be found to the problem when the piston is started impulsively and moves out of the tube with constant velocity. The solution consists of an expansion fan bounded by two regions of uniform flow and can be found exactly as in Section III.1. An alternative method of solution is to use the similarity property of the equations and boundary conditions to infer that u and a depend only on $\frac{x}{t}$. (See Exercise 4.)

If the piston moves into the fluid, the above solutions will break down when the characteristics intersect. No continuous solution then exists, and a discontinuity or "shock wave" must occur. This situation will be discussed in Chapter V.

4. TWO DIMENSIONAL STEADY FLOW

The next simple example of an inviscid compressible flow which we shall study will be two-dimensional steady flow. We shall assume that the flow is homentropic and irrotational. These assumptions are valid if, for example, the flow is initially uniform and everywhere continuous. Then, from the energy equation, the entropy is constant everywhere, and from Crocco's theorem, the vorticity is zero everywhere. The two-dimensional equations are, from (1.6) and (1.7),

$$\frac{\partial(\rho u)}{\partial x} + \frac{\partial(\rho v)}{\partial y} = 0, \tag{4.16}$$

$$u\frac{\partial u}{\partial x} + v\frac{\partial u}{\partial y} + \frac{a^2}{\rho}\frac{\partial \rho}{\partial x} = 0, \tag{4.17}$$

$$u\frac{\partial v}{\partial x} + v\frac{\partial v}{\partial y} + \frac{a^2}{\rho}\frac{\partial \rho}{\partial y} = 0. \tag{4.18}$$

If the flow is irrotational, we can define ϕ by $q = \text{grad}\,\phi$ and then eliminate ρ between equations (4.16), (4.17) and (4.18). This leads to the equation

$$(a^2-u^2)\frac{\partial^2\phi}{\partial x^2} - 2uv\frac{\partial^2\phi}{\partial x\partial y} + (a^2-v^2)\frac{\partial^2\phi}{\partial y^2} = 0. \tag{4.19}$$

Since the flow is irrotational and homentropic, Bernoulli's equation (4.5) holds everywhere and

$$a^2 = a_0^2 - \frac{(\gamma-1)}{2}(u^2+v^2). \tag{4.20}$$

Equation (4.19) can now be written in terms of ϕ and its derivatives; it is a quasi-linear second order equation for ϕ. To discuss (4.19) in detail requires knowledge of the theory of second order partial differential equations,† but a naive approach is possible, similar to that used in the previous section. Equation (4.19) may be written

$$(a^2-u^2)\frac{\partial u}{\partial x} - uv\left(\frac{\partial u}{\partial y} + \frac{\partial v}{\partial x}\right) + (a^2-v^2)\frac{\partial v}{\partial y} = 0, \tag{4.21}$$

and the irrotationality condition gives

$$\frac{\partial u}{\partial y} - \frac{\partial v}{\partial x} = 0. \tag{4.22}$$

Adding (4.21) to σ times (4.22) we obtain

$$\left((a^2-u^2)\frac{\partial}{\partial x} + (\sigma-uv)\frac{\partial}{\partial y}\right)u + \left((-\sigma-uv)\frac{\partial}{\partial x} + (a^2-v^2)\frac{\partial}{\partial y}\right)v = 0, \tag{4.23}$$

†A brief treatment is presented in the Appendix.

and the operators on u and v are identical if $\sigma = \pm a\sqrt{q^2-a^2}$, where $q^2 = u^2 + v^2$. Equation (4.23) then becomes

$$\left(\frac{\partial}{\partial x} + \left(\frac{-uv\pm a\sqrt{q^2-a^2}}{a^2 - u^2}\right)\frac{\partial}{\partial y}\right)\left(\int(a^2-u^2)du - \int(uv \pm a\sqrt{q^2-a^2})dv\right) = 0,$$

so that the characteristics are given by

$$\frac{dy}{dx} = \frac{-uv \pm a\sqrt{q^2-a^2}}{a^2 - u^2}, \tag{4.24}$$

and along these lines

$$(a^2-u^2)du - (uv \pm a\sqrt{q^2-a^2})dv = 0. \tag{4.25}$$

Thus the characteristics are real and the equations are hyperbolic only if $q > a$ and the flow is locally supersonic. When $q > a$, equation (4.24) can be rearranged in the form

$$a^2(dx^2 + dy^2) = (u\,dy - v\,dx)^2,$$

or

$$\pm a = u\frac{dy}{ds} - v\frac{dx}{ds},$$

where s is measured along a characteristic. Therefore the component of velocity normal to a characteristic is $\pm a$, and the angle between the characteristic and the velocity q at any point is $\pm\sin^{-1}(\frac{a}{q}) = \pm\sin^{-1}(\frac{1}{M}) = \pm\mu$, where μ is the local Mach angle. Thus, these two families of characteristics are identical with the Mach lines of the flow. The positive Mach lines are defined as those making an angle $+\mu$ with the streamline and the negative Mach lines make an angle $-\mu$, as shown in Figure 4.5.†

Along a characteristic, or Mach line, (4.25) holds. This relation can be transformed by putting $u = q\cos\theta$, $v = q\sin\theta$ to give

$$((M^2-1)^{\frac{1}{2}}dq \pm q\,d\theta)((M^2-1)^{\frac{1}{2}}\cos\theta \pm \sin\theta) = 0,$$

and

$$\frac{dq}{q} = \mp\frac{d\theta}{(M^2-1)^{\frac{1}{2}}}. \tag{4.26}$$

From Bernoulli's equation (4.20)

$$\frac{dq}{q} = \frac{dM}{M(1 + \frac{\gamma-1}{2}M^2)},$$

†Note that the positive sign in (4.24) gives the negative Mach line.

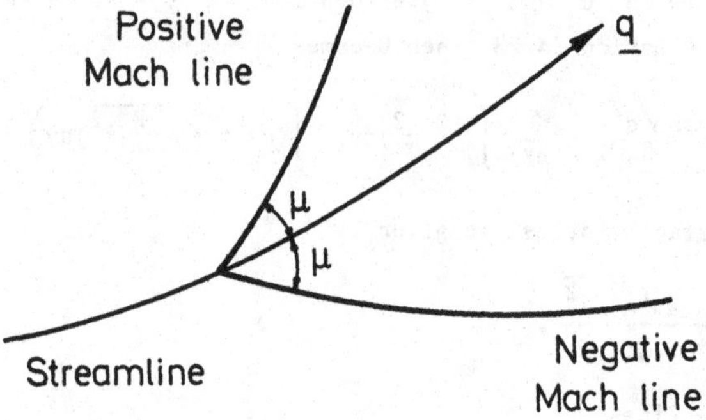

Figure 4.5. Mach lines.

and writing $M = \operatorname{cosec} \mu$, the relations which hold along the positive and negative characteristics reduce to

$$\theta \pm f(\mu) = \text{constant}, \tag{4.27}$$

where

$$f(\mu) = \mu + \frac{1}{\lambda} \tan^{-1}(\lambda \cot \mu) \tag{4.28}$$

and $\lambda^2 = (\gamma-1)/(\gamma+1)$.

Prandtl-Meyer Flow

As an example of two-dimensional steady flow, consider a uniform steady supersonic stream of Mach angle μ_1 flowing past a continuous convex corner which starts at 0, as illustrated in Figure 4.6. The influence

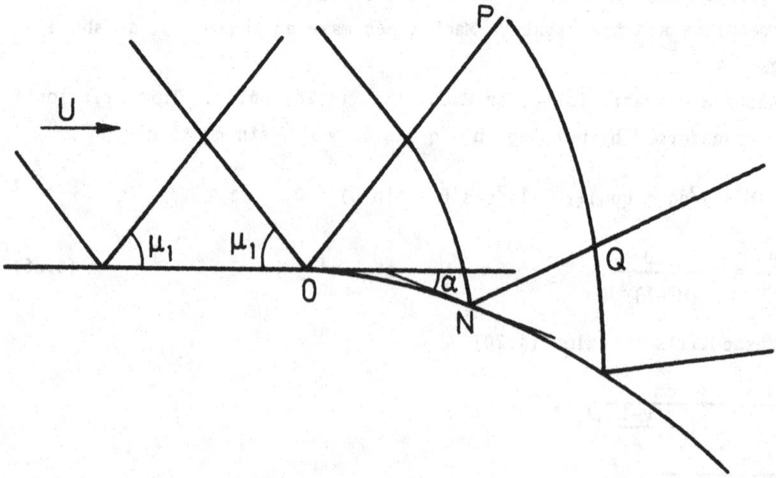

Figure 4.6. Supersonic flow expansion.

of the corner will be limited to the downstream side of the positive Mach line OP. We have used physical arguments to explain this fact in Section IV.1, but it may also be proved by applying the theory of "regions of influence" for hyperbolic partial differential equations to equation (4.19).† Upstream of the line OP, the conditions are uniform and the Mach lines (or characteristics) are straight lines, making angles $\pm\mu_1$ with the stream lines.

From the discussion in the previous section, we know that $\theta \mp f(\mu) =$ const. along the positive and negative Mach lines. Thus, on any negative Mach line which cuts OP,

$$\theta + f(\mu) = f(\mu_1),$$

and hence this relation holds throughout the flow. At a point N on the boundary the direction of the stream is known and $\theta = -\alpha$, so that at N, μ is given by $f(\mu) = f(\mu_1) + \alpha$. On the positive Mach line NQ,

$$\theta - f(\mu) = \text{constant} = -2\alpha - f(\mu_1),$$

and also

$$\theta + f(\mu) = f(\mu_1).$$

Thus at Q, $\theta = -\alpha$,

$$f(\mu) = f(\mu_1) + \alpha, \tag{4.29}$$

and θ and μ are constant along the Mach line NQ, which is therefore straight. Then, provided equation (4.29) can be solved for μ, the solution is known everywhere. This is another example of *simple wave flow* and is called a *Prandtl-Meyer* expansion.* From equation (4.28) the maximum value of $f(\mu)$ is $\pi/2\lambda$, which occurs when $\mu = 0$, and the minimum is $\pi/2$ when $\mu = \pi/2$. Therefore, the maximum angle through which the flow can be turned is $\frac{\pi}{2}(\frac{1}{\lambda} - 1)$. This can only be achieved if $\mu_1 = \pi/2$ and the original stream is sonic. If the angle turned through by the wall is larger than the maximum angle $\frac{\pi}{2\lambda} - f(\mu_1)$, then a region of zero pressure will form between the gas and the wall. In practice, viscosity becomes an important factor before the maximum angle is reached.

We now consider supersonic flow past a sharp corner as shown in Figure 4.7. Consideration of the characteristics leads to a result analogous to that found for shallow water theory and for the impulsively

†See Appendix.

*Note that Figure 4.6 is qualitatively identical with Figure 4.4 if it is inverted; 4.6 is drawn this way up by convention.

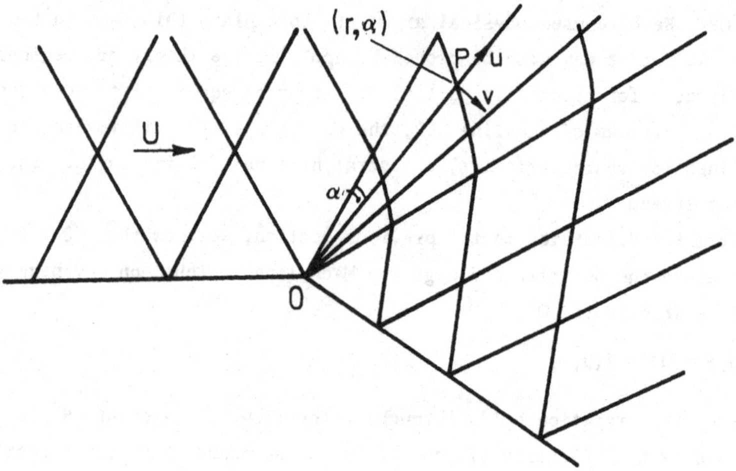

Figure 4.7. Supersonic flow past a corner.

started piston in the one-dimensional unsteady case. The flow consists of two regions of uniform flow separated by a fan-shaped region of expansion where the positive characteristics are all straight lines through the corner 0, as shown in Figure 4.7. This diagram is equivalent to that on p. 47 (but upside down). The explicit solution in the fan is now obtained by use of a similarity argument (as mentioned on p. 69).

Within the expansion fan, we use polar coordinates (r,α) with velocity components (u,v) as shown in Figure 4.7, where α is measured in a clockwise direction. Since there is no length scale in the problem the solution must depend on α alone. This means that conditions are constant along a positive characteristic, which is a ray of the fan. Using the fact that the velocity perpendicular to a characteristic is a, we see that the transverse velocity $v = a$. Bernoulli's equation (4.20) then becomes

$$u^2 + \frac{\gamma+1}{\gamma-1} a^2 = \text{constant} = V^2.$$

In view of this equation, we may put

$$u = V \sin \beta, \quad v = a = \lambda V \cos \beta, \tag{4.30}$$

where $\lambda^2 = (\gamma-1)/(\gamma+1)$. The condition for irrotational flow in these polar coordinates is

$$-v + \frac{du}{d\alpha} = 0,$$

and substituting from (4.30) gives

$$\lambda = \frac{d\beta}{d\alpha},$$

so that $\beta = \lambda\alpha + \text{constant}$. Now

$$\tan\beta = \frac{\lambda u}{v} = \frac{\lambda\sqrt{q^2-a^2}}{a} = \lambda\sqrt{M^2-1} = \lambda\cot\mu,$$

so that $\lambda\alpha + \text{constant} = \tan^{-1}(\lambda\cot\mu)$. Finally, when $\alpha = 0$, $\mu = \mu_1$ and

$$\lambda\alpha = \tan^{-1}(\lambda\cot\mu) - \tan^{-1}(\lambda\cot\mu_1).$$

This gives μ and then, using Bernoulli's equation, u and v at all points of the fan. This solution could also have been obtained by consideration of the characteristic equations.

If the corner is concave rather than convex, the characteristics will intersect and a continuous solution will no longer exist. A shock wave or discontinuity will then form. This case is considered in detail in the next chapter.

5. THE HODOGRAPH PLANE

One method of simplifying equation (4.19) for steady two dimensional irrotational flow is by using a Legendre transformation, which uses the velocity components u and v as independent variables; in these variables the equations become linear. The procedure is referred to as transforming from the physical to the *hodograph plane*. This idea has been used before in Section II.1 to simplify the incompressible flow problem with free boundaries. We write

$$u = \frac{\partial\phi}{\partial x},\quad v = \frac{\partial\phi}{\partial y},\quad f(u,v) = xu + yv - \phi,$$

whence

$$\frac{\partial f}{\partial u} = x + u\frac{\partial x}{\partial u} + v\frac{\partial y}{\partial u} - \frac{\partial\phi}{\partial u} = x,$$

and similarly $\partial f/\partial v = y$. Also,

$$1 = \frac{\partial}{\partial u}\left(\frac{\partial\phi}{\partial x}\right) = \frac{\partial^2\phi}{\partial x^2}\cdot\frac{\partial^2 f}{\partial u^2} + \frac{\partial^2\phi}{\partial x\partial y}\cdot\frac{\partial^2 f}{\partial u\partial v},$$

$$0 = \frac{\partial}{\partial v}\left(\frac{\partial\phi}{\partial x}\right) = \frac{\partial^2\phi}{\partial x^2}\cdot\frac{\partial^2 f}{\partial u\partial v} + \frac{\partial^2\phi}{\partial x\partial y}\cdot\frac{\partial^2 f}{\partial v^2},$$

and two further similar expressions may be written down. Hence

$$\frac{\partial^2 f}{\partial u^2} = D\,\frac{\partial^2 \phi}{\partial y^2}, \quad \frac{\partial^2 f}{\partial v^2} = D\,\frac{\partial^2 \phi}{\partial x^2}, \quad \frac{\partial^2 f}{\partial u \partial v} = -D\,\frac{\partial^2 \phi}{\partial x \partial y}$$

where

$$D = \begin{vmatrix} \dfrac{\partial^2 f}{\partial u^2} & \dfrac{\partial^2 f}{\partial u \partial v} \\[2ex] \dfrac{\partial^2 f}{\partial u \partial v} & \dfrac{\partial^2 f}{\partial v^2} \end{vmatrix} \quad \text{and} \quad \frac{1}{D} = \begin{vmatrix} \dfrac{\partial^2 \phi}{\partial x^2} & \dfrac{\partial^2 \phi}{\partial x \partial y} \\[2ex] \dfrac{\partial^2 \phi}{\partial x \partial y} & \dfrac{\partial^2 \phi}{\partial y^2} \end{vmatrix}.$$

This transformation is only valid if D is bounded and non-zero.† The equation (4.19) becomes

$$(a^2-u^2)\frac{\partial^2 f}{\partial v^2} + 2uv\,\frac{\partial^2 f}{\partial u \partial v} + (a^2-v^2)\frac{\partial^2 f}{\partial u^2} = 0, \tag{4.31}$$

where $a^2 = a_0^2 - \frac{(\gamma-1)}{2}(u^2+v^2)$ from Bernoulli's equation. The equation is now linear and may be transformed into an even simpler form by writing $u = q\cos\theta$ and $v = q\sin\theta$, so that equation (4.31) becomes

$$\frac{\partial^2 f}{\partial \theta^2} + \frac{q^2 a^2}{a^2-q^2}\,\frac{\partial^2 f}{\partial q^2} + q\,\frac{\partial f}{\partial q} = 0, \tag{4.32}$$

where a^2 is defined by (4.5). This is *Chaplygin's* equation.

Although the equation is now linear, it is not possible to transform the boundary conditions on solid boundaries into simple conditions in the hodograph plane, and the application of this method is limited. It has however proved possible to generate families of solutions of equation (4.32) by separating the variables and examining their form in the physical plane. For a detailed study of Chaplygin's equation see Schiffer [22].

EXERCISES

1. An inviscid, heat conducting gas is at rest in a uniform state $\rho = \rho_0$, $T = T_0$, $a = a_0$. Show that for small disturbances (acoustic waves) there exists a disturbance velocity potential which satisfies

$$(\gamma-1)\,C_v\,\frac{\partial T}{\partial t} = \frac{a_0^2}{\gamma}\,\nabla^2\phi - \frac{\partial^2 \phi}{\partial t^2}$$

†This excludes, for example, simple wave flow.

and

$$C_p \frac{\partial T}{\partial t} = \frac{\gamma k}{\rho_0} \nabla^2 T - a_0^2 \nabla^2 \phi,$$

where k is the constant conductivity and C_p and C_v are the specific heats at constant pressure and volume respectively.

Deduce that

$$\left(\frac{\partial^2}{\partial t^2} - \frac{a_0^2}{\gamma} \nabla^2\right)\left(\frac{k}{\rho_0} \nabla^2 \phi\right) - \left(\frac{\partial^2}{\partial t^2} - a_0^2 \nabla^2\right)\left(C_v \frac{\partial \phi}{\partial t}\right) = 0.$$

2. A pipe of length L is closed at one end, $x = 0$, and the other end is given by $x = L(1 + \varepsilon \cos t)$ where $\varepsilon \ll 1$. Assuming that the flow is one-dimensional, find u correct to $O(\varepsilon)$ if $\omega L \neq n\pi a_0$, where a_0 is the speed of sound in the undisturbed gas and n is any integer. Show also that

$$a = a_0 - \frac{\varepsilon(\gamma-1)}{2} L\omega \frac{\cos(\omega x/a_0)\cos \omega t}{\sin \omega L/a_0} + O(\varepsilon^2).$$

Comment on the case $\omega L = \pi a_0$.

3. A gas flows steadily from a large reservoir through a long convergent-divergent nozzle. The pressure of the gas in the reservoir is p_0. Describe in detail the series of changes that will occur in the flow in the nozzle if the pressure at the downstream end of the nozzle is slowly decreased from p_0 to zero.

4. A piston starts to move out of a semi-infinite pipe containing gas at rest at time $t = 0$. For $t > 0$ the piston moves with constant velocity U_0. Show that the flow consists of two uniform regions in the (x,t) plane, joined by an expansion fan for $-a_0 t < x < (\frac{(\gamma+1)}{2} U - a_0)t$. Show also that the solution in the fan depends only on the similarity variable x/t, and that the solution is

$$U = \frac{2}{\gamma+1}\left(a_0 + \frac{x}{t}\right), \quad a = \frac{2a_0}{\gamma+1} - \frac{(\gamma-1)}{(\gamma+1)} \frac{x}{t}.$$

5. A piston moves out of a semi-infinite pipe with speed ct for $t > 0$. Show that the gas velocity is given by

$$\gamma u = \left(a_0 + \frac{\gamma+1}{2} ct\right) - \left[\left(a_0 + \frac{\gamma+1}{2} ct\right)^2 - 2\gamma c(a_0 t + x)\right]^{\frac{1}{2}}.$$

If the piston moves *into* the pipe with speed ct show that a shock forms after a time $2a_0/((\gamma+1)c)$.

6. A steady two-dimensional supersonic uniform stream flows past a sharp corner at $x = 0$. Show that the maximum angle through which the flow can expand is $\pi/(2\lambda) - \pi/2$, where $\lambda^2 = (\gamma-1)/(\gamma+1)$. Show that this maximum angle can only occur if the original flow is sonic.

7. A uniform steady supersonic stream of Mach angle μ flows in the region $y > 0$, $x < 0$ bounded by a wall at $y = 0$. For $x > 0$, the equation of the wall is $y = \frac{1}{2}cx^2$. Obtain the equations of the Mach lines and show that a shock will form such that its furthest upstream point is at

$$y = x \tan \mu_1 = \frac{f'(\mu_1)\sin^2\mu_1}{c(f'(\mu_1)-1)},$$

where $f(\mu_1)$ is defined by (4.28).

8. Derive Chaplygin's equation (4.32) and show that the characteristics of the equation in the (u,v) plane are epicycloids.

Chapter V
Shock Waves

1. NORMAL SHOCK WAVES

In the previous chapter, we deferred consideration of the situation that occurs when a piston is pushed into a semi-infinite one-dimensional tube containing gas at rest. If the analysis used in Chapter IV when the piston is withdrawn is applied in this case, it leads to a solution which is not single valued and therefore not physically realistic. We first consider an example to clarify the situation.

Suppose that the piston moves into the tube with constant acceleration c so that the piston path is given by $x = -\frac{1}{2}ct^2$. It can be shown (Chapter IV, Exercise 5) that the velocity u in the gas is given by

$$\gamma u = a_0 - \frac{\gamma+1}{2}ct \pm \left[a_0^2 + (\gamma-1)a_0ct + \frac{(\gamma+1)^2}{4}c^2t^2 + 2\gamma cx\right]^{\frac{1}{2}} \qquad (5.1)$$

at a subsequent time $t > 0$. This solution is illustrated in Figure 5.1 for various values of t. The curves (i), (ii) and (iii) represent the velocity of the gas given by equation (5.1) when $t < 2a_0/(\gamma+1)c$, $t = 2a_0/(\gamma+1)c$ and $t > 2a_0/(\gamma+1)c$ respectively. From curve (iii) it can be seen that the solution becomes double valued when $t > 2a_0/(\gamma+1)c$ for a certain range of values of x.† This is not physically possible and in fact there will be a discontinuity, or *shock wave*, where the flow will change rapidly from the undisturbed state to a state with finite velocity.

The differential equations (4.11) and (4.12) for inviscid flow are not applicable at the shock, where the derivatives of the flow variables are infinite. If flows involving shock waves are to be described by the equations of inviscid flow, we must consider discontinuous solutions of

†Note the similarity between Figures 5.1 and 3.5.

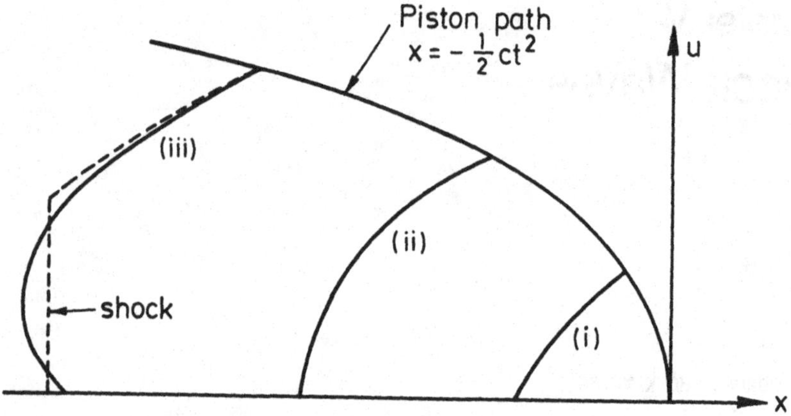

Figure 5.1. Solution of Equation (5.1).

these equations. There are many such solutions and to find the unique physically acceptable one it is necessary to invoke extra physical conditions to determine the position and strength of the shock. With this in mind, we now consider in detail the local behaviour of the gas near a shock wave.

There are several different approaches to the study of shock waves. The simplest physical argument considers a small region of the inviscid fluid near the shock and moving with the shock. The principles of conservation of mass, momentum and energy are applied to this region and the appropriate relations between conditions upstream and downstream of the shock can be derived. This approach is described by Becker [2] and Liepmann and Roshko [17]. The method has already been used in Chapter III.2 to study hydraulic jumps. A mathematical refinement of this method which has more general applicability uses conservation equations in integral form for the flow variables, instead of the differential equations which were derived in Chapter I. The advantage of this approach is that discontinuous solutions of the integral equations are mathematically possible. The method will be used here to treat one-dimensional unsteady shock waves.

We first derive the equations of one-dimensional unsteady flow using a slightly different approach from that used in Chapter I. We consider the fluid in a region D bounded by a curve C defined in the (x,t) plane by $x_1(t) < x < x_2(t)$ for $t_1 < t < t_2$, as shown in Figure 5.2.

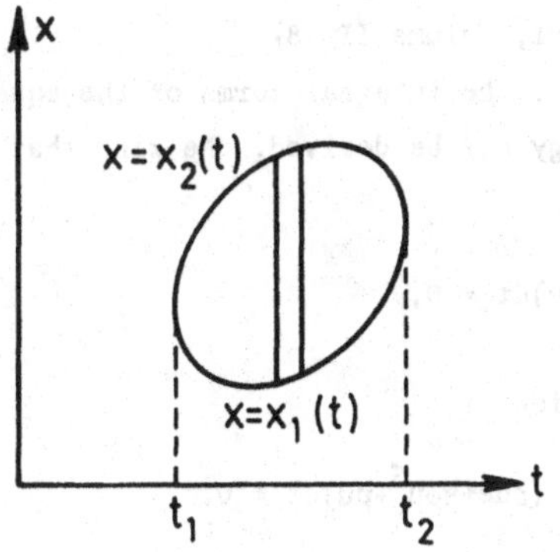

Figure 5.2. Regions of x,t space for conservation laws. Continuous flow.

In time $(t, t+\delta t)$, the amount of fluid which flows into the strip (x_1, x_2) in region D is

$$\left[\rho(x_1,t)\left(u(x_1,t) - \frac{dx_1}{dt}\right) - \rho(x_2,t)\left(u(x_2,t) - \frac{dx_2}{dt}\right)\right]\delta t.$$

When $t = t_1$ and t_2 there is no fluid in such a strip and

$$\int_{t_1}^{t_2}\left[(\rho(x_1,t)u(x_1,t) - \rho(x_2,t)u(x_2,t)) - \left(\rho(x_1,t)\frac{dx_1}{dt} - \rho(x_2,t)\frac{dx_2}{dt}\right)\right]dt = 0,$$

or

$$\int_C (\rho\ udt - \rho\ dx) = 0. \qquad (5.2)$$

If ρ and u are continuous, this equation may be transformed by Stokes' theorem to become

$$\iint_D \left(\frac{\partial\rho}{\partial t} + \frac{\partial}{\partial x}(\rho u)\right)dxdt = 0.$$

Thus (5.2) is equivalent to the equation of continuity (4.11) when there are no shock waves present. However, (5.2) is valid even if ρ and u are discontinuous at some points of C, and we have reformulated the equation of conservation of mass so that we can look for discontinuous solutions. A solution of (5.2) which is discontinuous is called a *weak solution* of (4.11). All solutions of the differential equation (4.11) satisfy the integral equation (5.2), but the reverse is *not* true. A rigorous treatment of weak solutions of partial differential equations is given in Courant and Hilbert, Volume II [8].

In a similar way, the integral forms of the equations of conservation of momentum and energy may be derived. We find that the momentum equation is

$$\int_C \rho u dx - (\rho u^2 + p)dt = 0, \tag{5.3}$$

and the energy equation is

$$\int_C (\rho e + \tfrac{1}{2}\rho u^2)dx - (\rho u e + \tfrac{1}{2}\rho u^3 + pu)dt = 0. \tag{5.4}$$

Both these equations and (5.2) are in the form

$$\int_C P dx - Q dt = 0,$$

and the corresponding differential equation

$$\frac{\partial P}{\partial t} + \frac{\partial Q}{\partial x} = 0,$$

which is satisfied by continuous differentiable solutions, is called a *conservation law*.

Now suppose Γ is a curve across which the flow variables are discontinuous. We choose C to be the rectangle $ABCD$ as shown in Figure 5.3. A and D are on opposite sides of the discontinuity at the point (x,t), and B and C are on opposite sides at $(x+\delta x, t+\delta t)$. Let the

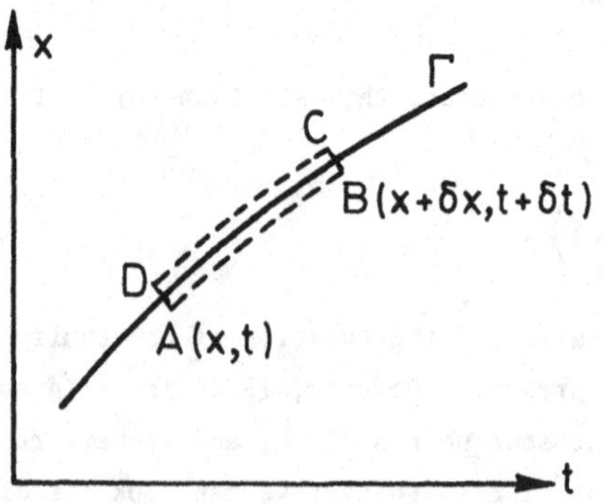

Figure 5.3. Regions of x,t space for conservation laws. Discontinuous flow.

values of the variables just in front of the shock be denoted by suffix 1 and those behind it by suffix 2. Then

$$\int_C Pdx - Qdt = \int_A^B + \int_B^C + \int_C^D + \int_D^A (Pdx - Qdt)$$

$$= (P_2\delta x - Q_2\delta t) + (-P_1\delta x + Q_1\delta t),$$

since $\int_B^C (Pdx - Qdt) \to 0$ and $\int_D^A (Pdx - Qdt) \to 0$ as $C \to B$ and $D \to A$. Thus in the limit as $\delta x, \delta t \to 0$,

$$\frac{dx}{dt} = \frac{Q_2-Q_1}{P_2-P_1} = \frac{[Q]}{[P]} . \tag{5.5}$$

From equations (5.2), (5.3) and (5.4) this gives

$$U = \frac{dx}{dt} = \frac{[\rho u]}{[\rho]},$$

$$= \frac{[\rho u^2+p]}{[\rho u]},$$

$$= \frac{[\rho ue+\frac{1}{2}\rho u^3+pu]}{[\rho e+\frac{1}{2}\rho u^2]},$$

where $U = dx/dt$ is the speed of the shock. These three equations can be rewritten as

$$\rho_1(U-u_1) = \rho_2(U-u_2), \tag{5.6}$$

$$p_1 + \rho_1(U-u_1)^2 = p_2 + \rho_2(U-u_2)^2, \tag{5.7}$$

$$h_1 + \tfrac{1}{2}(U-u_1)^2 = h_2 + \tfrac{1}{2}(U-u_2)^2, \tag{5.8}$$

where $h = C_pT = \gamma e$ and for an ideal gas, $h = \gamma p/(\gamma-1)\rho = a^2/(\gamma-1)$. They are called the *Rankine-Hugoniot* shock relations. It can be seen that it is the velocity relative to the shock which appears naturally in the shock relations. Indeed, equation (5.6) represents conservation of mass across the shock and equation (5.7) represents conservation of momentum. Equation (5.8) shows that the critical speed a_*, defined by (4.6), is constant across a stationary shock but not across a moving shock. The entropy jump can be calculated (see exercise 5) and shown to be non-zero and to depend on U. Hence for a genuinely unsteady shock, the flow behind the shock is not homentropic.†

†The same method can be applied to the *shallow water* equations (3.3) and (3.4) to obtain the jump conditions across a bore. Equation (3.4) is in conservation form and leads immediately to condition (3.10), but there is some difficulty over the correct conservation form for the other equation. (cont. on next page)

A weak solution of the equations of flow is given by two continuous solutions separated by a discontinuity across which the Rankine-Hugoniot relations hold. Weak solutions of conservation laws are not in general unique, however, and indeed it is found that equations (5.6), (5.7) and (5.8) are not sufficient to determine the flow pattern uniquely in a given situation. For example, if u_1, ρ_1, p_1, u_2 are given, there will be two possible solutions to the equations. We find that the physical condition that entropy must increase will usually ensure the existence of a unique discontinuous solution.

A third approach to the study of shock waves is motivated by the physical argument that discontinuities do not occur in a real gas but that in the neighborhood of a shock the variables change very rapidly, and some of the real gas effects which are negligible over most of the flow become important in this region of rapid change. The difference between the idealized shock (a) which is the discontinuous solution of the inviscid equations and the continuous shock (b) which occurs when real gas effects are taken into account is illustrated in Figure 5.4. The width of the shock wave, δ, is usually only of the order of a few mean free paths of the gas, and so approximation (a) is not unreasonable. To discuss the continuous solution (b) in the region of change we write $x = \delta X$ where $\delta \ll 1$, so that X is of order one within the shock, and then substitute in the viscous compressible equations of motion. It is then possible to find a continuous steady transition from one steady state to another,

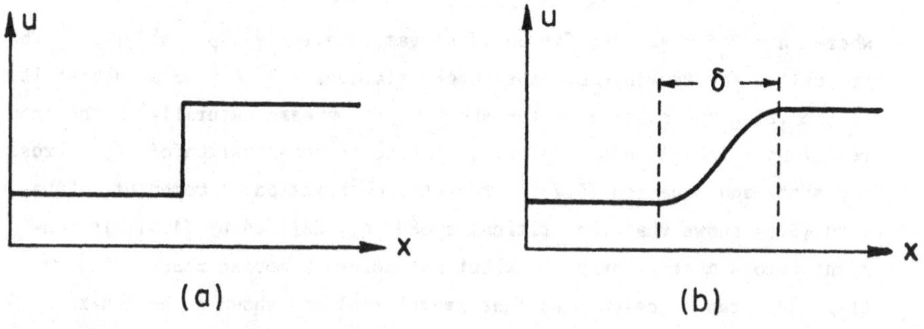

Figure 5.4. Shock profile.

†(footnote cont.)
For a momentum-conserving bore it is necessary to consider the *total* momentum in the x-direction, which leads to the conservation equation

$$\frac{\partial}{\partial t}(u\eta) + \frac{\partial}{\partial x}(u^2\eta + \tfrac{1}{2}g\eta^2) = 0.$$

Then, using the above method, equation (3.12) can be easily derived.

provided certain relations hold between the flow variables in the two states. These conditions turn out to be the Rankine-Hugoniot shock relations which have already been derived. The physical mechanism that allows this smooth transition may be viscosity or conductivity or a combination of both effects; the detailed analysis is described by Becker, Chapter 4 [2]. It can be shown that as the viscosity and conductivity approach zero, $\delta \to 0$ and the discontinuous solution, with the correct shock relations (5.6), (5.7), (5.8), is approached.

Finally, it is also possible, but more complicated, to introduce other real gas effects such as relaxation, dissociation or radiation which are capable of smoothing the discontinuities which occur in the inviscid model. This is also described briefly by Becker [2].

Equations (5.6), (5.7) and (5.8) can be manipulated so that the downstream variables are all written in terms of $M_1 = (U-u_1)/a_1$, which is the upstream Mach number relative to the shock. We assume that the shock travels from region 2 into region 1. We obtain

$$\frac{p_2}{p_1} = \frac{2\gamma M_1^2}{\gamma+1} - \frac{\gamma-1}{\gamma+1}, \tag{5.9}$$

$$\frac{\rho_2}{\rho_1} = \frac{\bar{u}_1}{\bar{u}_2} = \frac{(\gamma+1)M_1^2}{2+(\gamma-1)M_1^2}, \tag{5.10}$$

$$M_2^2 = \frac{2+(\gamma-1)M_1^2}{2\gamma M_1^2-(\gamma-1)}, \tag{5.11}$$

where $\bar{u}_1 = U-u_1$ and $\bar{u}_2 = U-u_2$ are the upstream and downstream velocities relative to the shock. Since the entropy cannot decrease we have

$$S_2 \geq S_1$$

or

$$\frac{p_2}{p_1} \geq \left(\frac{\rho_2}{\rho_1}\right)^\gamma.$$

In Figure 5.5, p_2/p_1, ρ_2/ρ_1, $\bar{u}_2/\bar{u}_1$, M_2^2 are plotted against M_1^2, and it can be seen that the condition that entropy increases can hold only if $M_1^2 \geq 1$. Thus the flow ahead of a shock must be supersonic relative to the shock. From Figure 5.5 it then follows that

$$p_2 \geq p_1, \tag{5.12}$$

$$\rho_2 \geq \rho_1, \tag{5.13}$$

$$\bar{u}_2 \leq \bar{u}_1, \tag{5.14}$$

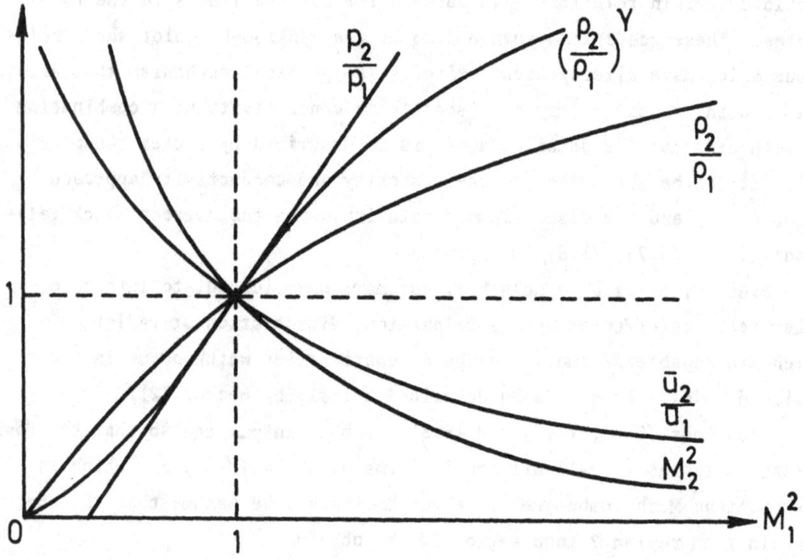

Figure 5.5. Normal shock relations.

and from (5.11) it can be seen that $M_2 \leq 1$ so that the flow behind a shock is subsonic relative to the shock.

A weak shock is defined by

$$\left|\frac{p_2-p_1}{p_1}\right| << 1$$

and occurs when M_1 is very close to 1. A strong shock occurs when $M_1 >> 1$, which implies that $p_2 >> p_1$. For such a shock

$$\frac{\bar{u}_1}{\bar{u}_2} = \frac{\rho_2}{\rho_1} \simeq \frac{\gamma+1}{\gamma-1} \tag{5.15}$$

and $M_2^2 \simeq (\gamma-1)/2\gamma$.

2. UNSTEADY SHOCK STRUCTURE

We have already mentioned that the structure of steady shock waves can be found by including real gas effects. As an example of unsteady shock structure, we shall consider the case of a weak viscous shock propagating into a gas at rest. In this case, the fluid velocities will be small everywhere and so from the discussion in the previous section it can be seen that the shock will travel with a speed which is close to a_0, the speed of sound in the undisturbed gas. Since the viscous terms are only important in the region of the shock, we can focus attention on this region and assume that the conditions at infinity both ahead of and behind the shock are uniform. Using approximations based on these observations, we shall find that the one-dimensional viscous unsteady equations reduce to a single second order partial differential equation for the velocity.

We first change to variables $y = x - a_0 t$ and t since we wish to examine in detail the flow in the region close to $y = 0$, which is the position of the shock in the limit as $\mu \to 0$. After some manipulation, equations (1.6), (1.41), (1.42) and (1.44) become

$$\rho_t + (u-a_0)\rho_y + \rho u_y = 0,$$
$$\rho u_t + \rho(u-a_0)u_y = -p_y + \tfrac{4}{3}\mu u_{yy},$$
$$p_t + (u-a_0)p_y + \gamma p u_y = \tfrac{4}{3}\mu(\gamma-1)u_y^2,$$

where we have assumed that there is no conductivity and also that μ is constant. Suppose that the shock width is $O(\delta)$ where $\delta \ll 1$. Then write $y = \delta Y$, since Y will be $O(1)$ within the shock. We also write $u = \bar{u}$, $\rho = \rho_0 + \bar{\rho}$, $p = p_0 + \bar{p}$ where $\bar{u}/a_0$, $\bar{\rho}/\rho_0$ and $\bar{p}/p_0$ are small quantities since we are considering a weak shock. The equations are then

$$\delta\bar{\rho}_t + (-a_0+\bar{u})\bar{\rho}_Y + (\rho_0+\bar{\rho})\bar{u}_Y = 0,$$
$$\delta(\rho_0+\bar{\rho})\bar{u}_t + (\rho_0+\bar{\rho})(-a_0+\bar{u})\bar{u}_Y + \bar{p}_Y = \frac{4}{3}\frac{\mu}{\delta}\bar{u}_{YY},$$
$$\delta\bar{p}_t + (-a_0+\bar{u})\bar{p}_Y + \gamma(p_0+\bar{p})\bar{u}_Y = \frac{4}{3}\frac{\mu}{\delta}(\gamma-1)\bar{u}_Y^2.$$

If we neglect all terms which are of the second order of small quantities we have

$$-a_0\bar{\rho}_Y + \rho_0\bar{u}_Y = 0,$$
$$-\rho_0 a_0\bar{u}_Y + \bar{p}_Y = 0,$$
$$-a_0\bar{p}_Y + \gamma p_0\bar{u}_Y = 0,$$

assuming that $\mu/\delta \ll 1$. These three equations are consistent and give

$$\bar{\rho} = \frac{\rho_0}{a_0}\bar{u} + \text{smaller terms} \tag{5.16}$$

and

$$\bar{p} = \rho_0 a_0\bar{u} + \text{smaller terms},$$

but do not determine $\bar{u}$. In this perturbation scheme it is necessary to retain higher order terms in the approximation in order to determine all the first order terms. We therefore keep the second order terms but neglect higher order powers of small quantities to give

$$-a_0\bar{\rho}_Y + \rho_0\bar{u}_Y = -\delta\bar{\rho}_t - \bar{u}\bar{\rho}_Y - \bar{\rho}\bar{u}_Y,$$

$$-\rho_0 a_0\bar{u}_Y + \bar{p}_Y = -\delta\rho_0\bar{u}_t - \rho_0\bar{u}\bar{u}_Y + a_0\bar{\rho}\bar{u}_Y + \frac{4}{3}\frac{\mu}{\delta}\bar{u}_{YY}, \tag{5.18}$$

$$-a_0\bar{p}_Y + \gamma p_0\bar{u}_Y = -\bar{p}_t - \bar{u}\bar{p}_Y - \gamma\bar{p}\bar{u}_Y. \tag{5.19}$$

On the right hand side of these equations we may substitute from (5.16) and (5.17), and then adding a_0 times (5.18) to (5.19) we have

$$0 = -2\rho_0\delta a_0\bar{u}_t + \frac{4}{3}a_0\frac{\mu}{\delta}\bar{u}_{YY} - \rho_0 a_0(\gamma+1)\bar{u}\bar{u}_Y. \tag{5.20}$$

In deriving (5.20) we have implicitly assumed that the terms are of comparable size, which implies certain conditions on the relative sizes of the variables and parameters. The exact nature of these assumptions is made clear by using a systematic perturbation scheme (see Exercise 9). We then find that the approximations used are valid only if the change in u across the shock is much greater than μ/ρ_0. The procedure used above to derive (5.20) is very similar to that used in Chapter III.3 to obtain the Korteweg-DeVries equation (3.23).

On transforming (5.20) back to unstretched variables we obtain

$$\frac{\partial u}{\partial t} + \frac{\gamma+1}{2}u\frac{\partial u}{\partial y} = \frac{2}{3}\frac{\mu}{\rho_0}\frac{\partial^2 u}{\partial y^2}. \tag{5.21}$$

This is *Burger's equation*, which is one of the few non-linear second order partial differential equations which can be solved exactly. The substitution

$$u = -\frac{8}{3(\gamma+1)\rho_0}\cdot\frac{1}{\theta}\frac{\partial\theta}{\partial y}$$

reduces (5.21) to the diffusion equation, which can be solved by transform methods. For simplicity we take as initial conditions when $t = 0$

$$u = u_1 \text{ if } y < 0 \quad\text{and}\quad u = 0 \text{ if } y > 0,$$

corresponding to a discontinuous shock wave at $y = 0$. The final solution for u can be shown to be

$$u = u_1\left[1 + \frac{e^{\kappa(y-\kappa\sigma t)}\operatorname{erfc}\left(-\frac{y}{2\sqrt{\sigma t}}\right)}{\operatorname{erfc}\left(\frac{y}{2\sqrt{\sigma t}} - \kappa\sqrt{\sigma t}\right)}\right]^{-1},$$

where

$$\sigma = \frac{2\mu}{3\rho_0}, \quad \kappa = \frac{\gamma+1}{4\sigma} u_1 \quad \text{and} \quad \text{erfc}(\alpha) = 1\text{-erf}(\alpha) = 1 - \frac{2}{\sqrt{\pi}} \int_0^{\alpha} e^{-\tau^2} d\tau.$$

As $t \to \infty$, keeping y finite,

$$u \to \frac{u_1}{1+\frac{1}{2}e^{\kappa(y-\kappa\sigma t)}} = \frac{u_1}{1+\frac{1}{2}e^{\kappa(x-(a_0+\frac{\gamma+1}{4}u_1)t)}},$$

which represents a wave with constant profile which is propagated with speed $(a_0 + \frac{\gamma+1}{4} u_1)$. The width of the wave is

$$O(\tfrac{1}{\kappa}) = O\left(\frac{\mu}{\rho_0 u_1}\right).$$

3. BLAST WAVES

One example involving an unsteady discontinuous shock that can be solved by using similarity methods is the *blast wave* which is caused by a violent explosion. An explosion is modelled as the sudden release of a large amount of energy E at a point O at time $t = 0$. This causes a strong spherical shock to form which then expands into the surrounding gas.

Suppose that before the explosion takes place, the gas is in equilibrium and the pressure and density are p_1 and ρ_1. Let the shock wave position be $r = R(t)$, the velocity of the gas just behind the shock be q, and the pressure and density behind the shock be p_2 and ρ_2. Assuming that the shock is strong, equation (5.10) gives

$$\frac{\dot{R}}{\dot{R}-q} = \frac{\rho_2}{\rho_1} = \frac{\gamma+1}{\gamma-1},$$

so that $q = 2\dot{R}/\gamma+1$. Also, p_1 will be negligible compared to p_2 so that equation (5.7) leads to

$$p_2 = \rho_1 \dot{R}^2 - \rho_2 (\dot{R}-q)^2 = \frac{2\rho_1 \dot{R}^2}{\gamma+1}.$$

The only physical quantities in this problem are r, t, E and ρ_1 (since p_1 is negligible), and the only non-dimensional combination of these quantities is

$$\lambda = r\left(\frac{\rho_1}{Et^2}\right)^{1/5}.$$

For a similarity solution, λ must be constant on the shock and hence

$$R \sim t^{2/5}.$$

The equations for the flow behind the shock reduce to ordinary differential equations in λ and the values of the variables on the shock are all known. One integral of the motion can be found immediately by observing that the total energy of the gas behind the shock remains constant and equals E. The other equations can be solved numerically.

A similar solution also exists in one or two dimensions. The problem is studied in more detail in Landau and Lifshitz [16] and Howarth [15].

4. OBLIQUE SHOCK WAVES

Shock Relations

We now consider the simple two-dimensional situation in which a uniform stream is incident on a plane shock at rest and making an acute angle α with the free stream direction. The shock conditions may be derived in an elementary way by resolving the flow into components normal and tangential to the shock and using the shock relations derived in the last section. Alternatively, the conditions may be derived more formally from the integral equations of motion. The continuity equation for two-dimensional steady flow may be written as

$$\int_C \rho u \, dy - \rho v \, dx = 0,$$

where C is any closed curve in the (x,y) plane. The momentum equation in the x direction is

$$\int_C (p+\rho u^2)dy - \rho uv \, dx = 0,$$

and a similar equation holds in the y direction. The integral form of the energy equation is

$$\int_C [pu + \rho u(\tfrac{1}{2}q^2+e)]dy - \int_C [pv + \rho v(\tfrac{1}{2}q^2+e)]dx = 0.$$

The analysis then proceeds exactly as for a one-dimensional unsteady shock. If we denote the velocities perpendicular to the shock by $u_1 = q_1 \sin\alpha$ and $u_2 = q_2 \sin(\alpha-\theta)$ and the velocities parallel to the shock by $v_1 = q_1 \cos\alpha$ and $v_2 = q_2 \cos(\alpha-\theta)$, as shown in Figure 5.6, then the shock relations are

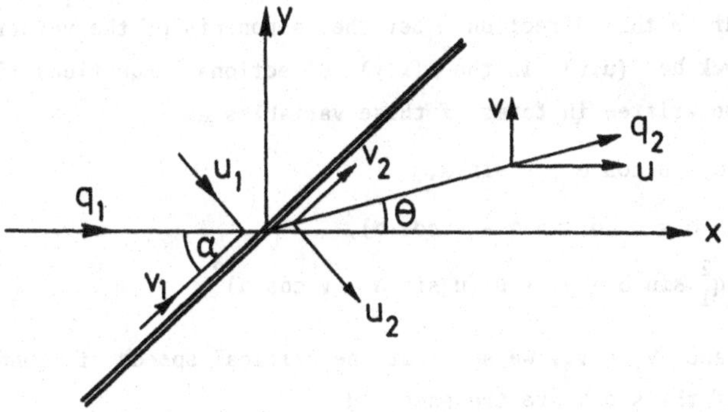

Figure 5.6. Oblique shock wave.

$$v_1 = v_2, \tag{5.22}$$

$$\rho_1 u_1 = \rho_2 u_2, \tag{5.23}$$

$$p_1 + \rho_1 u_1^2 = p_2 + \rho_2 u_2^2, \tag{5.24}$$

$$h_1 + \tfrac{1}{2}u_1^2 = h_2 + \tfrac{1}{2}u_2^2. \tag{5.25}$$

Thus the flow satisfies the Rankine-Hugoniot equations (5.6), (5.7) and (5.8) perpendicular to the shock, and parallel to the shock the velocity is continuous. In particular, inequality (5.14) shows that

$$q_1 \sin\alpha \geq q_2 \sin(\alpha-\theta),$$

but from (5.22) the velocity tangential to the shock is

$$q_1 \cos\alpha = q_2 \cos(\alpha-\theta),$$

and therefore $\tan\alpha \geq \tan(\alpha-\theta)$. This implies that θ is positive, and the stream is turned away from the normal to the shock. Thus a shock of this type provides a mechanism for deflecting a stream through a finite angle θ.

Since the flow in front of a one-dimensional shock must be supersonic, $q_1 \sin\alpha \geq a_1$, which implies that $q_1/a_1 \geq \operatorname{cosec}\alpha \geq 1$, and the flow upstream of a two-dimensional shock must be supersonic. However, downstream of an oblique shock the flow may be supersonic or subsonic.

<u>The Shock Polar</u>

It is useful to have a graphical method of representing solutions of the Rankine-Hugoniot equations. Coordinates x and y are defined as shown in Figure 5.6, with x parallel to the free stream q_1 and y perpendicular to this direction. Let the components of the velocity behind the shock be (u,v) in the (x,y) directions. Equations (5.22)-(5.24) may be written in terms of these variables as

$$\begin{aligned} q_1 \cos\alpha &= u\cos\alpha + v\sin\alpha, \\ \rho_1 q_1 \sin\alpha &= \rho_2(u\sin\alpha - v\cos\alpha), \\ p_1 + \rho_1 q_1^2 \sin^2\alpha &= p_2 + \rho_2(u\sin\alpha - v\cos\alpha)^2. \end{aligned} \tag{5.26}$$

From (5.25) and $v_1 = v_2$, we see that the critical speeds of sound on both sides of the shock are the same and

$$\begin{aligned} \frac{a_1^2}{\gamma-1} + \tfrac{1}{2}q_1^2\sin^2\alpha &= \frac{a_2^2}{\gamma-1} + \tfrac{1}{2}(u\sin\alpha - v\cos\alpha)^2 \\ &= \frac{\gamma+1}{2(\gamma-1)}a_*^2 - \tfrac{1}{2}q_1^2\cos^2\alpha. \end{aligned} \tag{5.27}$$

Eliminating p_1, ρ_1, p_2, ρ_2 from these equations we eventually obtain

$$v^2 = \frac{(q_1-u)^2(q_1u-a_*^2)}{\left((\frac{2}{\gamma+1})q_1^2 + a_*^2 - q_1u\right)}. \tag{5.28}$$

Thus, given q_1, this equation shows the relation between the possible values of u and v. The curve which this equation represents in the (u,v), or hodograph, plane for a given free stream defined by q_1 and a_* is called the *shock polar* and is sketched in Figure 5.7.

The inequality (5.14) implies that $u \le q_1$. Thus the closed loop of the graph is the only physically relevant solution curve.

For a point $P(u,v)$ on the polar, the line OP represents the velocity behind the shock in magnitude and direction. Thus we see that for a given deflection θ, there are two possible shocks and two possible values of the velocity behind the shock, represented by OP and OP'. For the flow past a sharp concave corner this gives a non-unique solution, although in practice the flow usually observed is the weaker shock represented by OP. Both shocks are possible, however, and the conditions imposed further downstream will determine which shock will occur.

From equation (5.26) we see that $\tan\alpha = (q_1-u)/v$, and in Figure 5.7 the angle PAO is $\pi/2 - \alpha$. Hence the shock which gives rise to velocity OP is perpendicular to the line PA and will therefore lie along ON. Thus P and P' correspond to quite different shocks, although the deflection is the same in both cases. The maximum deflection, θ_{max}, is the

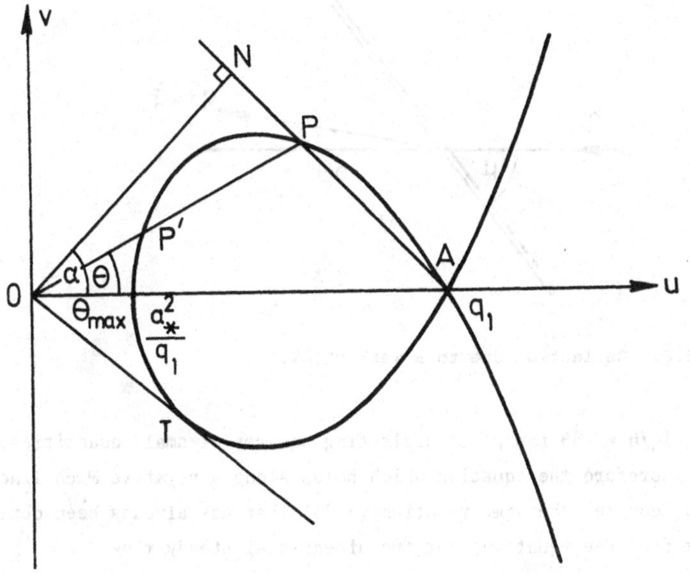

Figure 5.7. The shock polar.

angle between OA and the tangent from O to the polar; this is the greatest angle that the stream can be turned through by a single shock.

Weak Shocks

For a weak shock it follows from (5.9) that $M_1 \sin \alpha$ is near to unity. Writing (5.10) and (5.22) in terms of q_1, α, θ, q_2 and eliminating q_1/q_2 leads to

$$\tan \theta = \frac{(M_1^2 \sin^2\alpha - 1)\sin 2\alpha}{\gamma+1+(\gamma+\cos 2\alpha)(M_1^2 \sin^2\alpha-1)}$$

so that $\theta = O(M_1^2\sin^2\alpha-1)$ and the deflection is small. The point P will therefore be close to A and the line AN will approach the tangent to the shock polar at A. From (5.28), the slope of the tangent at A is $\pm\sqrt{q_1^2-a_*^2/a_*^2-(\gamma-1/\gamma+1)q_1^2}$, which can be written in terms of M_1 as $\pm(M_1^2-1)^{\frac{1}{2}}$ on using (5.27). Therefore the inclination of the corresponding shock is μ_1, and in the limit a weak shock approaches the Mach line. In the notation of Figure 5.8, for a very weak shock we use continuity parallel to the shock to obtain

$$q \cos \mu = (q+\delta q)\cos(\mu-\delta\theta),$$

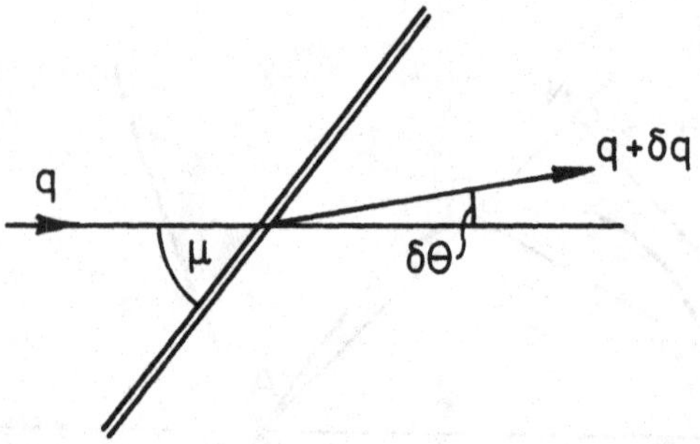

Figure 5.8. Deflection due to a weak shock.

so that $\delta q/q = -\delta\theta \tan\mu$ on neglecting squares of small quantities. This is therefore the equation which holds along a negative Mach line. It is, of course, the same relation (4.26) that has already been obtained directly from the equations for two-dimensional steady flow.

It is of particular interest to consider the change in entropy across a weak shock. Now

$$S_2 - S_1 = C_v \log\left\{\frac{p_2}{p_1}\left(\frac{\rho_1}{\rho_2}\right)^\gamma\right\},$$

and from (5.9) and (5.10)

$$\frac{p_2}{p_1}\left(\frac{\rho_1}{\rho_2}\right)^\gamma = \left[1 + \frac{2\gamma}{\gamma+1}(M_1^2\sin^2\alpha-1)\right]\left[\frac{1+(\gamma-1/\gamma+1)(M_1^2\sin^2\alpha-1)}{1+(M_1^2\sin^2\alpha-1)}\right]^\gamma$$
$$= 1 + O((M_1^2\sin^2\alpha-1)^3),$$

on expanding in powers of $(M_1^2\sin^2\alpha-1)$. Using the fact that $\theta = O(M_1^2\sin^2\alpha-1)$ from above, we see that the change in entropy $S_2 - S_1 = O(\theta^3)$. It is similarly possible to obtain the magnitude of the change in all the flow variables across a weak shock in terms of the deflection.

Shock Waves in Flows Past Bodies

We are now in a position to consider the flow of a uniform supersonic stream past a symmetrically placed two-dimensional wing with a pointed leading edge of angle 2θ. A plane shock will attach at the leading edge as shown in Figure 5.9(i) provided $\theta < \theta_{max}$, the maximum angle through which the free stream can be turned. When θ increases and becomes greater than θ_{max}, the shock becomes a curved cylindrical surface

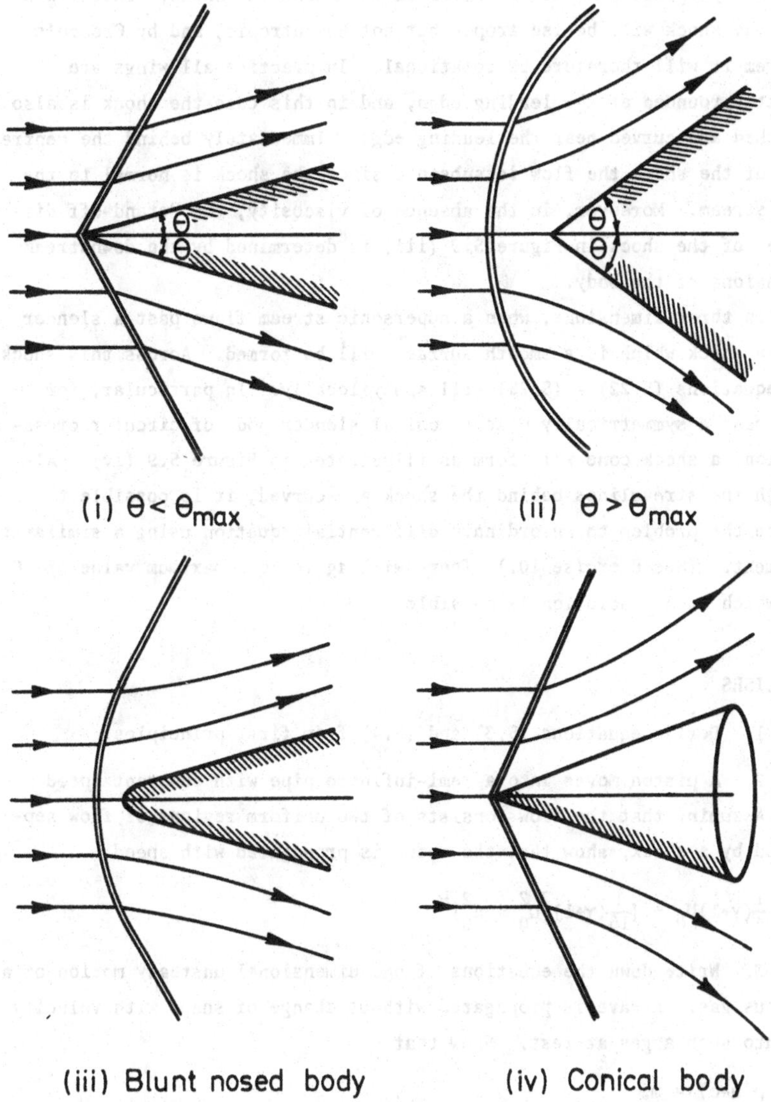

Figure 5.9. Supersonic Flow Past a Body.

and is detached from the leading edge as shown in Figure 5.9 (ii). Across a curved shock, the shock relations (5.22) - (5.25) will only apply locally. In particular, the varying slope of the shock will imply an entropy gradient perpendicular to the streamlines behind the shock. The flow behind the shock will be isentropic but not homentropic, and by Crocco's theorem it will therefore be rotational. In practice all wings are slightly rounded at the leading edge, and in this case the shock is also detached and curved near the leading edge. Immediately behind the centre part of the shock the flow is subsonic since the shock is normal to the free stream. Moreover, in the absence of viscosity, the "stand-off distance" of the shock in Figure 5.9 (iii) is determined by the downstream dimensions of the body.

In three dimensions, when a supersonic stream flows past a slender body a shock which is a smooth surface will be formed. Across this shock the equations (5.22) - (5.25) will apply locally. In particular, for flow past a symmetrically placed conical slender body of circular cross-section, a shock cone will form as illustrated in Figure 5.9 (iv). Although the streamlines behind the shock are curved, it is possible to reduce the problem to an ordinary differential equation using a similarity argument. (See Exercise 10.) There will again be a maximum value of θ for which such a solution is possible.

EXERCISES

1. Derive equations (5.3) and (5.4) from first principles.

2. A piston moves into a semi-infinite pipe with constant speed U_0. Assuming that the flow consists of two uniform regions of flow separated by a shock, show that the shock is propagated with speed

$$\frac{1}{4}(\gamma+1)U_0 + [\frac{1}{16}(\gamma+1)^2U_0^2 + a_0^2]^{\frac{1}{2}}.$$

3. Write down the equations of one-dimensional unsteady motion of a viscous gas. A wave is propagated without change of shape with velocity c into such a gas at rest. Show that

$$\rho(u-c) = m,$$

$$p + m(u-c) - \frac{4}{3}\mu\frac{du}{d\xi} = P,$$

$$\frac{m}{\gamma-1}\frac{p}{\rho} + P(u-c) - \tfrac{1}{2}m(u-c)^2 = Q,$$

where $\xi = x-ct$ and m, P and Q are constants. Hence derive the Rankine-Hugoniot shock relations for an inviscid gas.

4. Show that across a normal shock wave in a perfect gas

$$(\bar{u}_1-\bar{u}_2)^2 = (p_2-p_1)\left(\frac{1}{\rho_1} - \frac{1}{\rho_2}\right) \quad \text{and} \quad \frac{\rho_1}{\rho_2} = \frac{(\gamma+1)p_1 + (\gamma-1)p_2}{(\gamma-1)p_1 + (\gamma+1)p_2}.$$

A normal shock is reflected from a plane parallel rigid surface. If p is the pressure ratio for the reflected shock and π the pressure ratio for the incident shock, show that

$$p = \frac{(3\gamma-1)\pi-(\gamma-1)}{(\gamma-1)\pi+(\gamma+1)}.$$

Show that for a strong incident shock $p \sim (3\gamma-1)/(\gamma-1)$, whereas for a weak shock $p \sim \pi$.

5. Show that across a weak normal shock (i.e., for (M_1^2-1) small) the entropy jump is $(2\gamma(\gamma-1)/3(\gamma+1)^2)(M_1^2-1)^3 + O[(M_1^2-1)^4]$.

6. Show how the shock relations for an oblique shock can be derived from the integral equations of two-dimensional steady motion given in V.4.

7. A strong shock wave is defined by $q_1 \gg a_1$. Show that a_*^2 is approximately $\lambda^2 q_1^2$ and hence that the velocity components behind the shock approximately satisfy

$$\left(\frac{v}{u}\right)^2 = \left(\frac{q}{u} - 1\right)\left(1 - \frac{\lambda^2 q}{u}\right),$$

where $\lambda^2 = (\gamma-1)/(\gamma+1)$. Show that in this case the maximum deflection of the flow is approximately $\sin^{-1}(1/\gamma)$.

8. Two shocks intersect at a point P and have uniform supersonic flow ahead of them. By considering the intersection of the shock polars, show that the downstream conditions near P can be made compatible by a third shock through P and a line of slip flow, provided M_1 is large enough.

9. Derive equation (5.21) using a systematic perturbation scheme by writing

$$\rho = \rho_0(R_0 + \varepsilon R_1 + \ldots),$$
$$p = \rho_0 a_0^2(P_0 + \varepsilon P_1 + \ldots),$$
$$u = a_0(\varepsilon U_1 + \varepsilon^2 U_2 + \ldots),$$

$$y = \varepsilon L Y,$$

$$t = \frac{L}{a_0} T,$$

in the equations of V.2. ε is defined by the upstream boundary condition $u \to \varepsilon a_0$ as $y \to -\infty$ and L is an $O(1)$ length scale. Show that a necessary condition to derive Burger's equation is that $\varepsilon^2 = O(\mu / L a_0 \rho_0)$.

Show also that a solution of (5.21) depending only on the variable $(y-ct)$ is possible if $c = \frac{\gamma+1}{4} \varepsilon a_0$, and write down this solution.

10. A cone of semiangle α is placed in a supersonic stream U with its axis parallel to the stream. If u,v are the radial and transverse velocities behind the conical shock wave, show that

$$v = \frac{du}{d\theta} \quad \text{and} \quad \left(u + \frac{d^2u}{d\theta^2}\right)\left(1 - \frac{1}{a^2}\left(\frac{du}{d\theta}\right)^2\right) + \cot\theta \, \frac{du}{d\theta} + u = 0.$$

What are the boundary conditions on the shock?

Chapter VI
Approximate Solutions For Compressible Flow

1. LINEARIZED THEORY

The objective of this section is to find an approximation to the flow field when a small disturbance occurs in a uniform fluid motion. All the flow variables are assumed to differ by small amounts from their undisturbed values and the squares of these small quantities are assumed to be negligible.

Let the disturbed velocity be $q = (U+u,v,w)$ where $u,v,w << U$, the free stream velocity, and let the density $\rho = \rho_1 + \rho_2$, where $\rho_2 << \rho_1$ and ρ_1 is the density of the free stream. Since the undisturbed flow is uniform, the perturbed flow will be homentropic unless shock waves occur in the flow. However, any shocks that occur will be weak, and it was shown in Section V.4 that the entropy change across a weak shock is $O(\theta^3)$ where θ is the deflection caused by the shock. Since θ is small we can neglect the entropy change in our first order theory. Thus, if p_1 is the pressure in the free stream, the perturbed pressure $p = p_1(\rho/\rho_1)^\gamma = p_1(1+\gamma\rho_2/\rho_1)$ to the first order. The equations of motion may now be written in terms of the perturbed variables u,v,w,ρ_2, and second order quantities may be ignored. The equation of continuity becomes

$$\frac{\partial\rho_2}{\partial t} + U\frac{\partial\rho_2}{\partial x} + \rho_1\frac{\partial u}{\partial x} + \rho_1\frac{\partial v}{\partial y} + \rho_1\frac{\partial w}{\partial z} = 0, \tag{6.1}$$

and Euler's equations become

$$\frac{\partial u}{\partial t} + U\frac{\partial u}{\partial x} = -\frac{a_1^2}{\rho_1}\frac{\partial\rho_2}{\partial x}, \tag{6.2}$$

$$\frac{\partial v}{\partial t} + U\frac{\partial v}{\partial x} = -\frac{a_1^2}{\rho_1}\frac{\partial \rho_2}{\partial y}, \tag{6.3}$$

$$\frac{\partial w}{\partial t} + U\frac{\partial w}{\partial x} = -\frac{a_1^2}{\rho_1}\frac{\partial \rho_2}{\partial z}, \tag{6.4}$$

where a_1 is the speed of sound in the undisturbed flow. Defining $\zeta = \text{curl}\ \mathbf{q}$, we see that equations (6.2), (6.3) and (6.4) lead to

$$\frac{\partial \zeta}{\partial t} + U\frac{\partial \zeta}{\partial x} = 0,$$

so that $\zeta = f(x-Ut,y,z)$ and the vorticity is convected with the free stream velocity. Since the undisturbed flow is irrotational, the flow will be irrotational everywhere to first order. Thus we may define a perturbation potential ϕ by

$$\mathbf{q} = \text{grad}(Ux + \phi).$$

Substituting in terms of ϕ in equations (6.2), (6.3) and (6.4) and then integrating leads to

$$\frac{\partial \phi}{\partial t} + U\frac{\partial \phi}{\partial x} = -\frac{a_1^2}{\rho_1}\rho_2 = -\frac{(p-p_1)}{\rho_1}, \tag{6.5}$$

which is the linearized form of Bernoulli's equation. Substituting for ρ_2 from (6.5) into (6.1), we obtain

$$a_1^2\left(\frac{\partial^2\phi}{\partial x^2} + \frac{\partial^2\phi}{\partial y^2} + \frac{\partial^2\phi}{\partial z^2}\right) = \frac{\partial^2\phi}{\partial t^2} + 2U\frac{\partial^2\phi}{\partial x\partial t} + U^2\frac{\partial^2\phi}{\partial x^2}. \tag{6.6}$$

For steady flow this equation becomes

$$(1-M_1^2)\frac{\partial^2\phi}{\partial x^2} + \frac{\partial^2\phi}{\partial y^2} + \frac{\partial^2\phi}{\partial z^2} = 0, \tag{6.7}$$

where $M_1 = U/a_1$ is the Mach number of the free stream. For two-dimensional flow this equation can be derived immediately from (4.19). It is clear that equations (6.6) and (6.7) do not describe flows involving both supersonic and subsonic regions. These will occur when (M_1^2-1) is sufficiently small, and the flow is then said to be *transonic*. The linearized equations are also invalid for *hypersonic* flows when M_1^2 is very large. Both these exceptions will be discussed later in this chapter.

2. ACOUSTICS

The theory of acoustics deals with small disturbances in a gas at rest. It is therefore appropriate to use the linearized theory of the previous section with $U = 0$, on the assumption that $u,v,w \ll a_0$, the stagnation speed of sound. From (6.6), the acoustic wave equation is therefore

$$a_0^2 \nabla^2\phi = \frac{\partial^2\phi}{\partial t^2} . \tag{6.8}$$

This equation is consistent with the acoustic equation (4.4) which was derived previously.

For a point source of sound in an otherwise undisturbed gas, the resulting flow will have spherical symmetry, and the general solution of (6.8) in this case is

$$\phi = \frac{1}{r}\,(f(r-a_0t) + g(r+a_0t)),$$

where f and g are arbitrary functions. Since the disturbance originates at $r = 0$, the wave must be outgoing, and this *radiation condition* requires that $g = 0$. Then we can see that a spherical wave is propagated out from the origin with speed a_0 and that the potential decays like $1/r$. Also, an instantaneous disturbance at $t = 0$ will give rise to a wave front on $r = a_0t$ with no disturbance either ahead of or behind the front. In the case of cylindrical symmetry there is no simple solution of the equation, but it can be shown that at large values of r the potential due to a line source at $r = 0$ decays like $r^{-\frac{1}{2}}$. In this case the disturbance due to an instantaneous disturbance on $r = 0$ at $t = 0$ is not confined to the wave front at $r = a_0t$ but affects the whole region $r \leq a_0t$ for times $t > 0$. In the one-dimensional situation a plane wave is propagated without spatial decay, and an instantaneous initial disturbance gives rise to an instantaneous plane wave disturbance travelling with the speed of sound. This is the same effect that was mentioned on p. 64 when we considered the disturbance due to a point source of sound travelling with uniform supersonic velocity through a gas at rest. The problem in a fixed plane perpendicular to the direction in which the source is travelling is exactly equivalent to the above two dimensional problem with an instantaneous disturbance caused by the passage of the source through the plane in question. Thus we see that the flow is disturbed everywhere within the Mach cone. On the other hand, the disturbance due to a supersonic line source is equivalent to a one dimensional problem in a transverse plane, and so this disturbance is only felt on the Mach lines.

If the frequency of the sound wave is known, we may write

$$\phi = \mathrm{Re}[f(x,y,z)e^{i\omega t}],$$

so that f satisfies Helmholtz' equation

$$a_0^2\nabla^2 f = -\omega^2 f. \tag{6.9}$$

This equation can be simplified if ω is either large or small in comparison with the inverse of the appropriate time scale. When ω is small we can write

$$f = f_0 + \omega^2 f_1 + \omega^4 f_2 + \dots,$$

where the f_i are solutions of Poisson's equation. When ω is large we write

$$f = e^{i\omega(\psi_0+(1/\omega)\psi_1+\dots)}$$

as in the W.K.B. approximation for ordinary differential equations. The leading term ψ_0 therefore satisfies

$$\left(\frac{\partial\psi_0}{\partial x}\right)^2 + \left(\frac{\partial\psi_0}{\partial y}\right)^2 + \left(\frac{\partial\psi_0}{\partial z}\right)^2 = \frac{1}{a_0^2}, \tag{6.10}$$

which is the *Eikonal equation*. This is a first order nonlinear partial differential equation and may be solved by characteristic methods. However, since (6.10) is of lower order than (6.9) it may not be possible to apply the boundary conditions directly to solutions of (6.10). The study of the Eikonal equation forms the basis of *geometrical optics*; further reference to this subject can be found in Officer [20] or Landau and Lifschitz [16].

3. THIN WING THEORY

The equations of the upper and lower surfaces of a thin two-dimensional wing may be defined by $y = f_\pm(x)$ for $0 < x < c$, where $f_\pm(0) = 0$ and $f'_\pm(x)$ is small. Then on the body the boundary condition that the velocity is tangential is

$$f'_\pm(x) = \frac{\partial\phi/\partial y}{U+\partial\phi/\partial x} \quad \text{on} \quad y = f_\pm(x),$$

which may be approximated to give

$$\frac{\partial \phi}{\partial y} = Uf'_{\pm}(x) \quad \text{on} \quad y = \pm 0 \tag{6.11}$$

in linearized theory. We now consider separately the cases of subsonic and supersonic flow past a thin wing.

Subsonic Flow

In subsonic two-dimensional steady flow we write $1-M_1^2 = \beta^2$, and equation (6.7) becomes

$$\beta^2 \frac{\partial^2 \phi}{\partial x^2} + \frac{\partial^2 \phi}{\partial y^2} = 0. \tag{6.12}$$

The problem of solving this equation subject to condition (6.11) may be transformed into an incompressible flow problem past a "similar" body by a suitable stretching of the variables. Let the coordinates for the incompressible problem be (X,Y), the potential be Φ and the body be given by $Y = f_{\pm}(X)$. Then the incompressible problem for flow past the same thin wing is

$$\frac{\partial^2 \Phi}{\partial X^2} + \frac{\partial^2 \Phi}{\partial Y^2} = 0,$$

with

$$\frac{\partial \Phi}{\partial Y} = Uf'_{\pm}(X) \quad \text{on} \quad Y = \pm 0,$$

where we have assumed that the linearization of the boundary condition is again valid. Writing $\phi = \beta^{\lambda}\Phi$, $x = X$, $y = \beta^{\nu}Y$, we see that the two problems are identical if $\lambda = \nu = -1$. The problem of incompressible flow past thin wings is discussed as an application of complex variable methods in the next chapter. The solution of the compressible problem can be found from that of the corresponding incompressible problem, and it is possible to relate various features of the two flows. From the linearized Bernoulli equation (6.5)

$$\frac{p-p_0}{\rho_0} = -U \frac{\partial \phi}{\partial x} \quad \text{for compressible flow,}$$

and

$$\frac{p-p_0}{\rho_0} = -U \frac{\partial \Phi}{\partial X} = -U\beta \frac{\partial \phi}{\partial x} \quad \text{for incompressible flow,}$$

and hence

$$(C_p)_{inc} = \beta(C_p)_{comp},$$

where $C_p = p-p_0/\frac{1}{2}\rho_0U^2$ is the *pressure coefficient*. This is the Prandtl-Glauert similarity rule. It can also be shown that a similarity relation exists between two different subsonic flows past the same thin wing. One result of this similarity is that since the drag on a body in inviscid incompressible flow without circulation is zero, the drag on a thin wing in subsonic flow without circulation is also zero.

Supersonic Flow

For supersonic flow the linearized equation is

$$B^2 \frac{\partial^2\phi}{\partial x^2} = \frac{\partial^2\phi}{\partial y^2} \tag{6.13}$$

where $B^2 = M_1^2-1$. This equation is hyperbolic, the characteristics, or Mach lines, are the straight lines $x \pm By =$ constant, and the general solution of (6.13) is

$$\phi = F(x-By) + G(x+By).$$

The linearized boundary condition on the body is given by (6.11). Suppose that $\phi = \phi_\pm$ in $y > 0$ and $y < 0$ respectively. Since the flow is supersonic, $\phi_+ \equiv 0$ for $y > x/B$ and

$$\phi_+ = \begin{cases} F(x-By), & \text{for } x > By, \\ 0 \quad , & \text{for } x < By. \end{cases}$$

Applying condition (6.11),

$$\left(\frac{\partial\phi_+}{\partial y}\right)_{y=0} = -BF'(x) = Uf_+'(x) \quad \text{for} \quad 0 < x < c,$$

where c is the length of the body. Thus

$$\phi_+ = -\frac{U}{B} f_+(x-By) \quad \text{for} \quad 0 < x-By < c, \tag{6.14}$$

and similarly

$$\phi_- = \frac{U}{B} f_-(x+By) \quad \text{for} \quad 0 < x+By < c. \tag{6.15}$$

The leading Mach lines $x \pm By = 0$ correspond to weak shocks. For the symmetric case $f_+(c) = f_-(c) = 0$, this solution is sketched in Figure 6.1 (i). If $f_+(c) = f_-(c) \neq 0$, then we may define new axes Ox', Oy' along and perpendicular to the *chord* line† of the body as shown in Figure 6.1

†For a thin wing with pointed ends the chord line is naturally taken through the two pointed ends. If, as is often the case, the wing has a cusped trailing edge as in Figure 6.2, then the chord line is normally taken through the cusp in such a direction that there is zero lift on the wing when the external stream is parallel to the chord line.

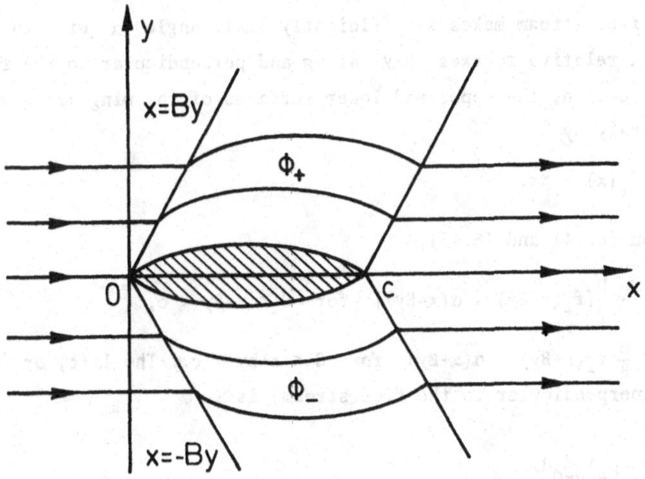

(i) Supersonic stream past symmetric body

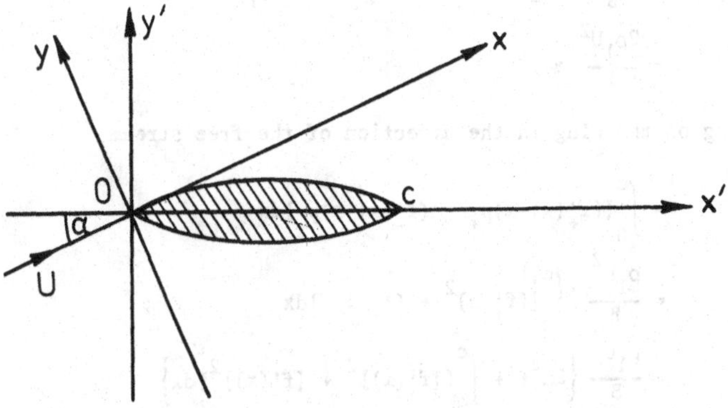

(ii) Supersonic stream past body at incidence

Figure 6.1. Supersonic stream past thin symmetrical wing at incidence.

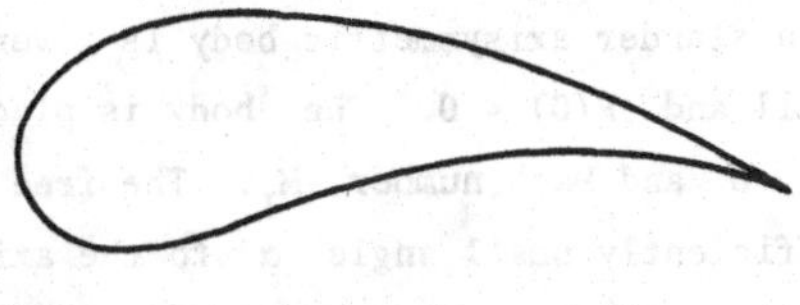

Figure 6.2. Aerofoil with cusped trailing edge.

(ii), so that the wing is given by $y'_{\pm} = f(x')$, and $f_{\pm}(c) = 0$. Suppose that the free stream makes a sufficiently small angle α with the axis Ox'. Then relative to axes Oxy along and perpendicular to the free stream direction, the upper and lower surfaces of the wing are given approximately by

$$y = f_{\pm}(x) - \alpha x.$$

Then, from (6.14) and (6.15),

$$\phi_{+} = -\frac{U}{B}[f_{+}(x-By) - \alpha(x-By)] \quad \text{for} \quad 0 < x-By < c,$$

and $\phi_{-} = \frac{U}{B} f_{-}(x+By) - \alpha(x+By)$ for $0 < x+by < c$. The lift, or force on the wing perpendicular to the free stream, is

$$\begin{aligned}
&\int_0^c (p_{-}-p_{+})_{y=0}\,dx \\
&\quad = \rho_1 U \int_0^c \left(\frac{\partial\phi_{+}}{\partial x} - \frac{\partial\phi_{-}}{\partial x}\right)dx \qquad \text{(from (5.5))} \\
&\quad = \frac{\rho_1 U^2}{B}\,[-f_{+}(x) + \alpha x - f_{-}(x) + \alpha x]_0^c \\
&\quad = \frac{2\rho_1 U^2}{B}\,\alpha c.
\end{aligned}$$

The drag on the wing in the direction of the free stream

$$\begin{aligned}
&= \int_0^c [(f'_{+}(x)-\alpha)p_{+} - (f'_{-}(x)-\alpha)p_{-}]\,dx \\
&= \frac{\rho_1 U^2}{B} \int_0^c [(f'_{+}-\alpha)^2 + (f'_{-}-\alpha)^2]\,dx \\
&= \frac{\rho_1 U^2}{B}\left\{2\alpha^2 f + \int_0^c ([f'_{+}(x)]^2 + [f'_{-}(x)]^2)\,dx\right\}.
\end{aligned}$$

Thus the drag is always positive, in contrast to the subsonic result. This drag is called the *wave drag* and is related to the energy propagated away from the wing between the leading and trailing Mach lines.

4. SLENDER BODY THEORY

The surface of a slender axisymmetric body is given by $r = R(x)$, where $R'(x)$ is small and $R(0) = 0$. The body is placed in a uniform stream with velocity U and Mach number M_1. The free stream direction is inclined at a sufficiently small angle α to the axis Ox so that the linearized approximation will be valid. We define cylindrical polar

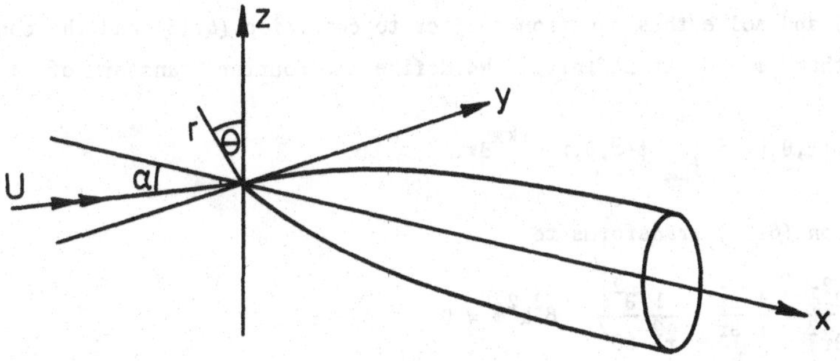

Figure 6.3. Flow past a slender body.

coordinates (r,θ,x) as shown in Figure 6.3 so that the free stream direction lies in the (x,z) plane. Then the free stream velocity will be

$(U \sin \alpha \cos \theta, -U \sin \alpha \sin \theta, U \cos \alpha)$ in (r,θ,x) directions.

The normal to the body is in the direction $(1,0,-R'(x))$, and since the velocity must be perpendicular to the normal on the body,

$$U \sin \alpha \cos \theta + \frac{\partial \phi}{\partial r} - R'(x)(U \cos \alpha + \frac{\partial \phi}{\partial x}) = 0 \quad \text{on} \quad r = R(x).$$

Neglecting second order terms, this becomes

$$\frac{\partial \phi}{\partial r} = UR'(x) - U\alpha \cos \theta \text{ on } r = R(x). \tag{6.16}$$

On writing the linearized equation for steady flow (5.7) in terms of (r,θ,x), we have

$$(1-M_1^2)\frac{\partial^2 \phi}{\partial x^2} + \frac{\partial^2 \phi}{\partial r^2} + \frac{1}{r}\frac{\partial \phi}{\partial r} + \frac{1}{r^2}\frac{\partial^2 \phi}{\partial \theta^2} = 0. \tag{6.17}$$

The form of this equation suggests that ϕ will be singular at $r = 0$, and so in (6.16), $\partial\phi/\partial r$ must be evaluated on $r = R$ and not at $r = 0$.

It is convenient to decompose ϕ into two parts by writing $\phi = \phi_0 + \phi_1$, where ϕ_0 and ϕ_1 both satisfy (6.16) but

$$\frac{\partial \phi_0}{\partial r} = UR'(x), \quad \frac{\partial \phi_1}{\partial r} = -U\alpha \cos \theta \quad \text{on} \quad r = R(x).$$

Thus $\phi_0 = \phi_0(r,x)$ is the potential when the body is placed symmetrically in the stream.

Both the subsonic and supersonic axisymmetric flows can be solved by integral transform methods. Since the techniques are similar we shall consider in detail only the subsonic case. We first write $1-M_1^2 = \beta^2$ in (6.17) and solve this equation subject to condition (6.16) and the condition that $\phi \to 0$ at infinity. We define the Fourier Transform of ϕ by,

$$\bar{\phi}(r,\theta,k) = \int_{-\infty}^{\infty} \phi(r,\theta,x)e^{ikx}dx.$$

Equation (6.17) transforms to

$$\frac{\partial^2\bar{\phi}}{\partial r^2} + \frac{1}{r}\frac{\partial\bar{\phi}}{\partial r} + \frac{1}{r^2}\frac{\partial^2\bar{\phi}}{\partial\theta^2} - \beta^2k^2\bar{\phi} = 0$$

and can be solved by separation of variables. Writing $\bar{\phi} = f(r)g(\theta)$ leads to

$$g(\theta) = A_n \cos n\theta + B_n \sin n\theta,$$

where n must be an integer for single-valued solutions, and $f(r)$ satisfies

$$r^2 \frac{d^2f}{dr^2} + r\frac{df}{dr} - (\beta^2k^2r^2+n^2)f = 0.$$

This is Bessel's equation with an imaginary argument, and the solution which tends to zero as $r \to \infty$ is

$$f = K_n(\beta|k|r).$$

(See Erdelyi [9].) From boundary conditions (6.16) we see that it is only necessary to consider $\bar{\phi}$ in the form

$$\bar{\phi} = A_0(k)K_0(\beta|k|r) + A_1(k)K_1(\beta|k|r)\cos\theta. \tag{6.18}$$

From the asymptotic properties of Bessel functions (Murray [19]),

$$K_0(z) \sim -\log z + O(1)$$

and

$$K_1(z) \sim \frac{1}{z} + O(1) \quad \text{as} \quad z \to 0,$$

so that, as $r \to 0$, assuming k to be finite,

$$\bar{\phi} \sim -A_0(k)\log r + \frac{A_1(k)\cos\theta}{\beta|k|r} + O(1).$$

Thus as $r \to 0$, $\phi \sim -a_0(x)\log r + a_1(x)\cos\theta/r$, where $a_0(x)$ is the inverse transform of $A_0(k)$ and $a_1(x)$ is the inverse transform of $A_1(k)/\beta|k|$. Condition (6.16) gives

$$a_0(x) = -UR(x)R'(x) \tag{6.19}$$

and

$$a_1(x) = U\alpha[R(x)]^2.$$

It is now necessary to invert formula (6.18). Since the inverse transform of $K_0(\beta|k|r)$ is $\frac{1}{2}(x^2+\beta^2r^2)^{-\frac{1}{2}}$ (see Erdelyi, [10]), by (6.19) and the convolution theorem the first term in (6.18) inverts to give

$$\phi_0 = -\frac{U}{2}\int_0^c \frac{R(s)R'(s)ds}{((x-s)^2+\beta^2r^2)^{\frac{1}{2}}},$$

which is the potential due to flow past a body at zero incidence. The second term in (6.18) can also be inverted by similar methods so that the final solution is

$$\phi = -\frac{U}{2}\int_0^c \frac{R(s)R'(s)ds}{((x-s)^2+\beta^2r^2)^{\frac{1}{2}}} - \frac{U}{2}\,\alpha\cos\theta\,\frac{\partial}{\partial r}\left\{\int_0^c \frac{(R(s))^2ds}{((x-s)^2+\beta^2r^2)^{\frac{1}{2}}}\right\}. \quad (6.20)$$

The potential ϕ_0 represents a distribution of sources of strength $-\frac{1}{2}UR(x)R'(x)$ along the x-axis from 0 to c and can alternatively be derived by solving the integral equation obtained by assuming a source distribution of this form. The potential ϕ_1 may similarly be represented by a line distribution of doublets. This method is described in Liepmann and Roshko [17].

The solution to the supersonic problem may be found similarly by using a Laplace transform in x instead of a Fourier transform. Alternatively, the body may be represented by a distribution of sources along the x-axis. From Figure 6.4 it can be seen that the flow at P will be affected only by the part of the body upstream of A, where P lies on

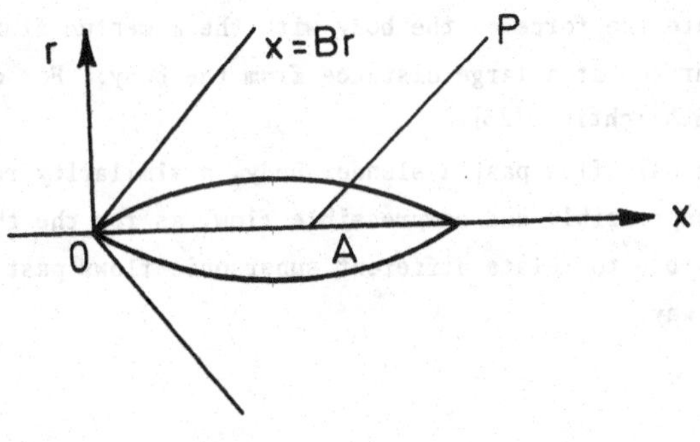

Figure 6.4. Supersonic flow past a slender body.

the Mach cone through A. For a wing at zero incidence in supersonic flow, we assume a source distribution $m(x)$ along the x axis from $x = 0$ to c. The potential at P will then be

$$\phi = \int_0^{x-Br} \frac{m(s)ds}{((x-s)^2-B^2r^2)^{\frac{1}{2}}},$$

where the limit $x-Br$ is appropriate since sources further downstream could not influence conditions at P. To apply the boundary condition we need to evaluate $\partial\phi/\partial r$ on $r = R(x)$, where R is small compared to c. First put $x-s = Br \cosh \theta$, so that

$$\phi = \int_0^{\cosh^{-1}(x/Br)} m(x - Br \cosh \theta)d\theta$$

$$\sim m(x)\cosh^{-1} \frac{x}{Br} \quad \text{as} \quad r \to 0$$

$$\sim m(x) \log r.$$

Therefore $\partial\phi/\partial r \sim m(x)/r$ as $r \to 0$, and using (6.16) gives $m(x) = -UR(x)R'(x)$. Thus

$$\phi = -U \int_0^{x-Br} \frac{R(s)R'(s)ds}{((x-s)^2-B^2r^2)^{\frac{1}{2}}}, \tag{6.21}$$

where $R(s) = 0$ if $s > c$. The region of influence of the body in this case extends over the whole fluid downstream of the Mach cone $x = Br$. This is in contrast to the two-dimensional result where the flow is affected only between the Mach lines at the leading and trailing edges of the body.

For both the above problems, the lift and drag on the bodies may be found by integrating the pressure forces after using Bernoulli's equation. This is usually a complicated procedure, and it is sometimes simpler to equate the force on the body with the momentum flux through a "control" surface at a large distance from the body. For details of this method see Schlichting [23].

For subsonic flow past a slender body, a similarity rule exists between incompressible and compressible flow, as for the thin wing. It is also possible to relate different supersonic flows past similar bodies in the same way.

5. LIMITATIONS OF LINEARIZED FLOW

We have already pointed out that the linearized equations of flow do not hold when M_1^2 is near to unity or when M_1^2 is very large. In this section we consider these two cases in more detail, show why the linear approximation does not apply, and derive an appropriate approximate equation for each situation. The linearized equations also become invalid for the flow far from the body for any value of M_1. Even though the quantities neglected in deriving the equations are always small, the cumulative effect of these small discrepancies will become appreciable at a sufficient distance from the body. It is possible to derive a uniformly valid approximation which holds everywhere in flow field; this is described by Van Dyke [25].

Transonic Flow

In transonic flow, the coefficient (M_1^2-1) of $\partial^2\phi/\partial x^2$ in the linear equation (6.7) is small, and we are not justified in neglecting terms such as $\partial\phi/\partial x \cdot \partial^2\phi/\partial x^2$, $\partial\phi/\partial y \cdot \partial^2\phi/\partial x^2$ compared with $(M_1^2-1)\partial^2\phi/\partial x^2$ in deriving equation (6.7). To derive a valid small disturbance approximation for steady two-dimensional transonic flow it is simpler to start from equation (4.19). The linear approximations used in Section VI.1 are still valid except in the coefficient of $\partial^2\phi/\partial x^2$, and equation (4.19) becomes

$$\left(a^2 - (U + \frac{\partial\phi}{\partial x})^2\right)\frac{\partial^2\phi}{\partial x^2} + a_1^2\frac{\partial^2\phi}{\partial y^2} = 0.$$

But

$$\begin{aligned}\frac{a^2}{a_1^2} &= (\frac{\rho}{\rho_1})^{\gamma-1} = \left(1 + \frac{\rho_2}{\rho_1}\right)^{\gamma-1} \\ &= 1 + (\gamma-1)\frac{\rho_2}{\rho_1} + \ldots \\ &= 1 + (\gamma-1)\left(-\frac{U}{a_1^2}\frac{\partial\phi}{\partial x}\right) + O((\frac{\partial\phi}{\partial x})^2) \ .\end{aligned}$$

and

$$a^2 - (U+\partial\phi/\partial x)^2 = a_1^2(1-M_1^2 - (\gamma+1)U/a_1^2\ \partial\phi/\partial x) + O((\partial\phi/\partial x)^2).$$

Thus the equation which is valid for small disturbances in transonic flow is

$$(1-M_1^2)\frac{\partial^2\phi}{\partial x^2} + \frac{\partial^2\phi}{\partial y^2} = \frac{(\gamma+1)U}{a_1^2}\frac{\partial\phi}{\partial x}\frac{\partial^2\phi}{\partial x^2} \ . \tag{6.22}$$

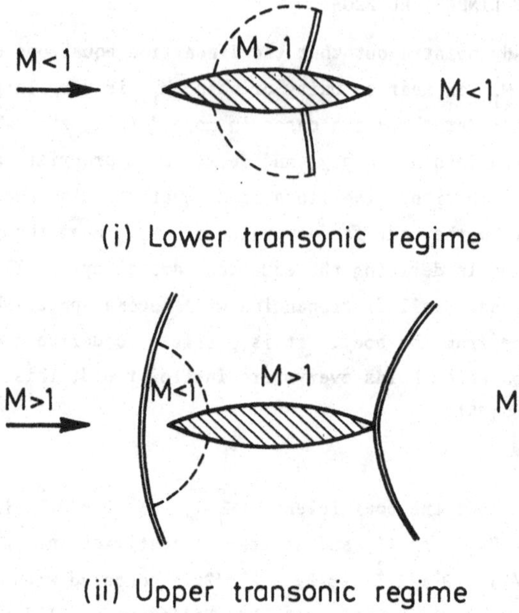

Figure 6.5. Transonic flow past a thin wing.

This equation is non-linear, and its qualitative properties are less well understood than in corresponding subsonic and supersonic cases. It changes character from hyperbolic to elliptic depending on the sign of $(1-M_1^2 - (\gamma+1)U/a_1^2\ \partial\phi/\partial x)$ and is said to be of mixed type. It is, however, possible to find similarity laws between different transonic flows past similar thin wings of the type found for subsonic flow in Section VI.2. (See Exercise 5.) For further details of the flow in the transonic regime see Liepmann and Roshko, [17] or Landau and Lifschitz, [16]. Some possible types of transonic flow past a wing are shown in Figure 6.5.

Hypersonic Theory

The linearized approximations of Section VI.1 are no longer valid when M_1 is so large that $M_1\delta = O(1)$, where δ is the maximum inclination of the body to the free stream. This means that for a slender body in a uniform stream of Mach number M_1 the slope of the Mach wave at the leading edge of the body is comparable with the slope of the body. The leading Mach line is no longer a weak shock in this situation, and it is necessary to consider the shock relations to determine the relative magnitudes of the various flow quantities between the shock and the body.

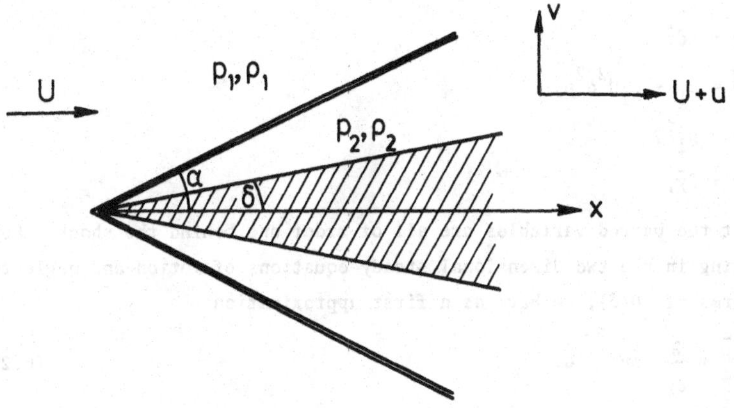

Figure 6.6. Hypersonic flow.

We consider the two-dimensional steady case illustrated in Figure 6.6. On the body, the boundary condition gives

$$\frac{v}{u+U} = O(\delta).$$

Therefore, $v/U = O(\delta)$ behind the shock, since $u \ll U$. Considering the flow parallel to the shock when the shock makes an angle α with the free stream,

$$\frac{u}{U} = -\frac{v}{U}\tan\alpha = O(\alpha\delta),$$

assuming that α is also small. Then, using the Rankine-Hugoniot equations (5.23) and (5.24) perpendicular to the shock, we have

$$\frac{\rho_2}{\rho_1} = \frac{U\sin\alpha}{(U+u)\sin\alpha - v\cos\alpha} = O(1) \tag{6.23}$$

and

$$\frac{p_2-p_1}{\rho_1 U^2} = \frac{\rho_1 U\sin\alpha}{\rho_1 U^2}[U\sin\alpha - (U+u)\sin\alpha + v\cos\alpha] = O(\alpha\delta). \tag{6.29}$$

The third equation (5.25) leads to the relation

$$\alpha = \frac{\gamma-1}{2\gamma}\delta - \frac{1}{\gamma^2 M_1^2} + \text{smaller terms},$$

and hence $\alpha = O(\delta)$. To consider the flow behind the shock, we use the above estimates to write

$$u = U\delta^2\bar{u},$$

$$v = U\delta\bar{v},$$

$$p = p_1 + \rho_1 U^2\delta^2\bar{p},$$

$$\rho = \rho_1\bar{\rho},$$

$$y = \delta\bar{y},$$

so that the barred variables are all of order one behind the shock. Substituting in the two-dimensional steady equations of motion and neglecting terms of $O(\delta)$, we have as a first approximation

$$\frac{\partial\bar{\rho}}{\partial x} + \frac{\partial}{\partial\bar{y}}(\bar{\rho}\bar{v}) = 0, \tag{6.25}$$

$$\frac{\partial\bar{u}}{\partial x} + \bar{v}\frac{\partial\bar{u}}{\partial\bar{y}} = -\frac{1}{\bar{\rho}}\frac{\partial\bar{p}}{\partial x}, \tag{6.26}$$

$$\frac{\partial\bar{v}}{\partial x} + \bar{v}\frac{\partial\bar{v}}{\partial\bar{y}} = -\frac{1}{\bar{\rho}}\frac{\partial\bar{p}}{\partial\bar{y}}. \tag{6.27}$$

The flow behind the shock will no longer be irrotational or homentropic since the shock is not weak, and we therefore also need the energy equation, which is approximately

$$\left(\frac{\partial}{\partial x} + \bar{v}\frac{\partial}{\partial\bar{y}}\right)\left(\frac{\bar{p}}{\bar{\rho}^\gamma}\right) = 0. \tag{6.28}$$

Equations (6.25), (6.27) and (6.28) do not depend on $\bar{u}$ and are exactly equivalent, with suitable change of notation, to the one-dimensional unsteady equations which were studied in Chapter IV. The approximate boundary condition for the flow past a body given by $y = f(x)$ is

$$\bar{v} = f'(x) \quad \text{on} \quad \bar{y} = f(x),$$

and the problem is mathematically identical with the flow caused by a piston in a tube moving with velocity $f'(t)$. If this problem can be solved for $\bar{\rho}$, $\bar{v}$, $\bar{p}$ we can then find $\bar{u}$ from equation (6.26). A similar analogue also exists between three-dimensional steady and two-dimensional unsteady hypersonic small disturbance flows.

One further simplification which can be used to advantage in the above equations is the *Newtonian* approximation, where $(\gamma-1) = \varepsilon$ is assumed to be small. To the first order in ε, the ratio of the densities across the shock is then $O(\varepsilon)$. The thin layer of very dense gas between the shock and the body is called a *shock layer*, and the structure of this layer can be determined explicitly.

If M_1 is so large that $M_1\delta >> 1$ then the shock is a *strong* shock and the shock relations simplify. In the particular case of a power law body given by $y = Ax^k$, a similarity solution can be found which depends only on the variable $\bar{y}/x^k$. This similarity solution is in some ways analogous to the blast wave solution considered in Chapter V. For further details of hypersonic flow see Cole, [6] and Hayes and Probstein, [14].

EXERCISES

1. A uniform stream U flows past a thin two-dimensional aerofoil with Mach number M_1 less than one. Show that the pressure perturbation on the aerofoil is $(1-M_1^2)^{-\frac{1}{2}}$ of the pressure perturbation for incompressible flow of a stream U past the same aerofoil. Show that the circulation is $(1-M_1^2)^{-\frac{1}{2}}$ of that in the incompressible case and verify that Kutta- Joukowski formula for the lift on the aerofoil is unaltered.

2. A thin two-dimensional aerofoil is in a supersonic stream M_1 and the upper and lower surfaces of the aerofoil are defined by $y = f_+(x)$ and $y = f_-(x)$ for $0 < x < c$. Show that the forces on the wing have a moment about the leading edge $x = 0$ given by

$$\frac{\rho U^2}{M_1^2-1}\left\{c^2\alpha + \int_0^c (f_+(x) + f_-(x))dx\right\},$$

where α is the angle of incidence of the wing. What simple design criterion would you use to avoid angular instability of the wing?

3. Use Laplace transforms to solve equations (6.17) subject to (6.16) in the supersonic case when $\alpha = 0$.

If $R(0) = 0$ and $R(c) = a$, show that the drag on the body is $\pi(p_0 - p_B)a^2$ to first order, where p_0 is the pressure in the free stream and p_B is the pressure on the base $x = c$. Explain how to obtain (but do not calculate) the drag when $a = 0$.

[(i) The solution of $f''(r) + 1/r\, f'(r) - \lambda^2 f(r) = 0$ which is bounded as $r \to \infty$ is $K_0(\lambda r)$,

(ii) $K_0(\lambda r) \sim -\log(\lambda r) + O(1)$ as $r \to 0$,

(iii) The Laplace Transform of $H(x-\beta)/(x^2-\beta^2)^{\frac{1}{2}}$ with respect to x is $K_0(\beta p)$.]

4. A circular cone of small semi-vertical angle δ is placed in a supersonic stream, Mach number M_1, pressure p_1, density ρ_1 and velocity U, at a small angle of attack α. Show that the pressure p on the surface of the cone is given by

$$p-p_1 = \rho_1 U^2[\delta^2\log(\frac{2}{\beta\delta}) - \tfrac{1}{2}(\delta^2+\alpha^2) - 2\alpha\delta\cos\theta + \alpha^2\cos 2\theta],$$

where $\beta = (M_1^2-1)^{\frac{1}{2}}$ and θ is the azimuthal angle.

5. The perturbation potential ϕ for steady two-dimensional transonic flow past a thin wing satisfies equation (6.22). Show that the transformation

$$\phi = \frac{(1-M'^2)^{\frac{1}{2}}}{(1-M^2)^{\frac{1}{2}}}\frac{U}{U'}\frac{\tau}{\tau'}\phi'$$

relates the transonic flow past a thin wing of thickness ratio τ to that past an affinely related thin wing of thickness ratio τ', provided that

$$\frac{\tau}{\tau'} = \frac{M'^2(1-M^2)^{3/2}}{M^2(1-M'^2)^{3/2}}.$$

Chapter VII
Complex Variable Methods

1. STEADY FLOWS PAST TWO DIMENSIONAL BODIES

The use of a complex potential w to describe incompressible inviscid flows past bodies is discussed in many elementary texts (see for example Milne-Thomson [18]) and will not be repeated here. The class of boundary shapes for which explicit solutions can be found is limited and the methods are indirect. Moreover, many of them may be of little practical value, since small viscous effects would result in flow separation from the body and a dramatic alteration of the streamline pattern. The objective of this chapter is to develop methods which are relevant to problems of flows with small viscosity. We shall restrict attention to free streamline flows and flows past thin wings.

It is however valuable to recall Blasius' Theorem concerning the forces on such a body, in the form

$$X-iY = \frac{i\rho}{2}\int_C \left(\frac{dw}{dz}\right)^2 dz, \tag{7.1}$$

where C is the body cross-section, and to examine in detail the problem of steady flow past a flat plate at incidence α_0, assuming that the flow is attached as in Fig. 7.1 (i). The solution is obtained by observing that the conformal transformation $z = \zeta + \frac{1}{\zeta}$ transforms the region exterior to $y = 0$, $-2 < x < 2$ into the region exterior to the circle $|\zeta| = 1$, leaving unaltered conditions far from the plate. Thus the problem reduces to solving for the flow past a circle, and it is easily verified that

$$w = Ue^{-i\alpha_0}\zeta + \frac{Ue^{+i\alpha_0}}{\zeta} \tag{7.2}$$

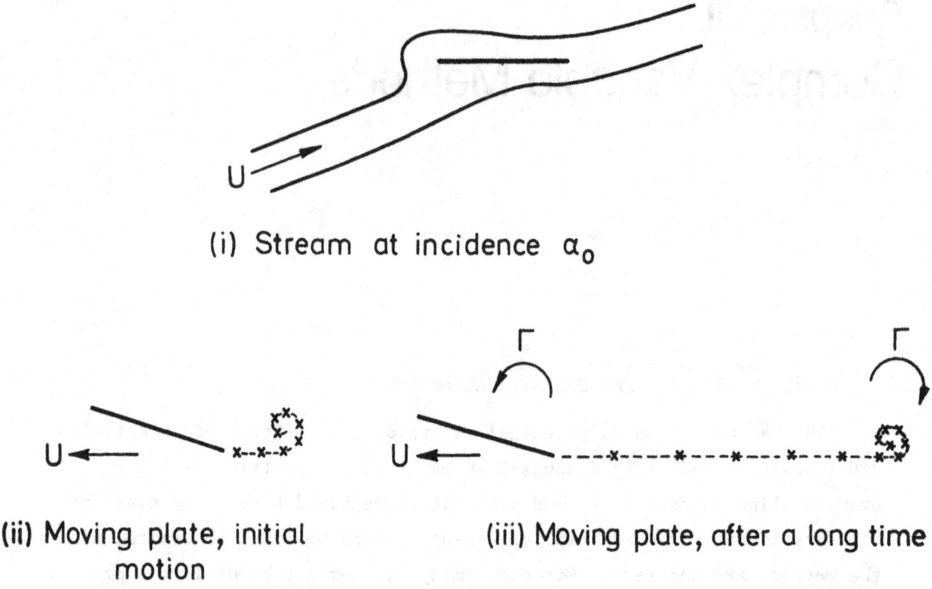

Figure 7.1. Flow past a flat plate.

has a streamline $|\zeta| = 1$ and represents a stream at angle α_0 to the real axis as $|\zeta| \to \infty$.

Now using Blasius' Theorem (7.1),

$$X-iY = \frac{i\rho}{2}\int_{\Sigma}\left(Ue^{-i\alpha_0} - \frac{Ue^{i\alpha_0}}{\zeta^2}\right)^2 \frac{\zeta^2 d\zeta}{\zeta^2-1},$$

where Σ is the circle at infinity. Hence $X-iY$ is zero, and there are no forces on the plate.

To obtain a solution which models the physical phenomenon of lift, we note that (7.2) is not the unique solution: a term $(i\Gamma/2\pi)\log\zeta$ may be included which still satisfies the boundary conditions. On using Blasius' Theorem we obtain $X = 0$, $Y = \rho U\Gamma$, and there is a lift force which is made possible by the existence of circulation Γ round the plate. If we consider the flat plate moving at incidence into a fluid at rest, circulation is still necessary for there to be lift. However Kelvin's Theorem (1.14) implies that the circulation round a large contour enclosing the plate will be zero if the plate, and hence the fluid, started from rest. Thus an equal, but opposite sense, circulation must exist in the fluid and be created by the motion of the plate; that is, the plate sheds vorticity behind it as in Fig. 7.1 (ii). A detailed description of the phenomenon of vortex shedding may be found in Batchelor pp. 438-441 [1]. The crucial result is that a long time after the plate started from rest, most of the vorticity is concentrated in a "starting" vortex as in Fig. 7.1 (iii), with the vorticity decaying to zero elsewhere behind the body. For this reason no further singularities need be introduced into the flow problem to model the vorticity which has been shed.

To determine the value of the circulation Γ we observe that dw/dz is singular at the leading and trailing edges $z = \pm 2$. No choice of Γ will remove both singularities, and we adopt the *Kutta condition* that the velocity will be bounded at the trailing edge, so that $\Gamma = 4\pi U \sin \alpha_0$ and $Y = 4\pi\rho U^2 \sin \alpha_0$. A complete discussion of the Kutta condition at the trailing edge involves considering the viscous flow in the boundary layer adjacent to the plate and is beyond the scope of this book. A simple argument valid for any cusped trailing edge is that the solution for an inviscid flow turning a sharp corner is given by $w = z^n$, where the flow expands by turning through a positive angle if $n < 1$, and z is measured from the corner. Near the corner there will be large velocities and hence, from Bernoulli's equation, the pressure will drop to zero. This implies that the flow separates, and it seems reasonable to assert that viscous effects will ensure that it separates smoothly. A similar argument applies at a sharp leading edge, but in this case the flow reattaches with only a small separated region where $p \equiv 0$, which is neglected.

This solution assumes that the flow is attached at all points of the body where the slope is continuous, and is unlikely to occur in practice unless α_0 is reasonably small. For large values of α_0, for flow past bluff bodies in general and for flow through orifices, flow separation will occur and the fluid region will be bounded by a combination of fixed and free boundaries as already described in Chapter II.1. Some explicit solutions may be obtained by transforming to hodograph plane variables, and we shall describe two examples in detail. The hodograph plane is defined by $Q = L + i\theta = \log(U/(dw/dz))$, where U is a suitable reference velocity, and Q will be an analytic function of w. As shown in Chapter II.1 the fluid occupies fixed regions in the Q and w planes, and the problem is solved if a conformal transformation can be found which maps one region onto the other. From the Riemann mapping theorem, such a transformation will be unique if (i) the boundary in the Q plane is mapped onto the boundary in the w plane, (ii) three points in the Q plane (including possibly the point at infinity) are mapped onto three given points in the w plane, and (iii) the interior maps into the interior. The Schwarz-Christoffel theorem gives the conformal transformation which maps a polygon onto the upper half plane. Thus the procedure is to map the given regions in both the Q and w planes onto the upper half plane of an intermediary variable. For details of the theorems and more examples, reference may be made to Milne-Thomson [18] and Birkhoff and Zarantonello [3].

Example. Two-Dimensional Jet.

Consider the jet formed through a slit of width $2h$ in a semi-infinite container of fluid whose stagnation pressure is greater than that in the atmosphere. Let the container be $y > 0$, the slit be $y = 0$, $-h < x < h$, and the jet be $y < 0$, $-f(y) < x < f(y)$, where $f(0) = h$ and $f(\infty) = H$ as in Figure 7.2 (i). On the jet boundary, from Bernoulli's equation with no external body forces, q is constant, say equal to U, so that the rate of volume flow in the jet is $2UH$ and the jet boundaries are the streamlines $\psi = \pm UH$. The value of U is determined from $p_0 = p_a + \frac{1}{2}\rho U^2$, where p_0 is the stagnation pressure in the fluid and p_a the pressure in the atmosphere. On the jet boundary, since $L = \log U/q$, we have $L = 0$, and on the container wall $\theta = 0$ or $-\pi$. The fluid therefore occupies the region $-UH < \psi < UH$ in the w plane, and $-\pi < \theta < 0$, $L > 0$ in the Q plane, with corresponding points as shown in Figures 7.2 (i), (ii), (iii). In this example the appropriate transformations are

$$\zeta = \cosh Q \quad \text{and} \quad \zeta = -i \exp(-\frac{\pi w}{2HU}),$$

which can either be spotted or found using the Schwarz-Christoffel result. In the ζ plane, B is -1, B' is +1 and C, C', F are at the origin so that the uniqueness is assured by fixing the points A, B and C in the w plane and observing that FE is interior to both domains.

Now $dw/dz = Ue^{-Q}$ by definition, and in principle we can eliminate Q and ζ and hence integrate for $w(z)$. In practice this can be complicated, and if we are only interested in the free boundary shape the analysis can be simplified. On the free boundary B'C', $Q = i\theta$ and $w = \phi - iUH$ so that

$$-\frac{\pi\phi}{2HU} + i\frac{\pi}{2} = \log(i \cos \theta) = \log \cos \theta + i\frac{\pi}{2}$$

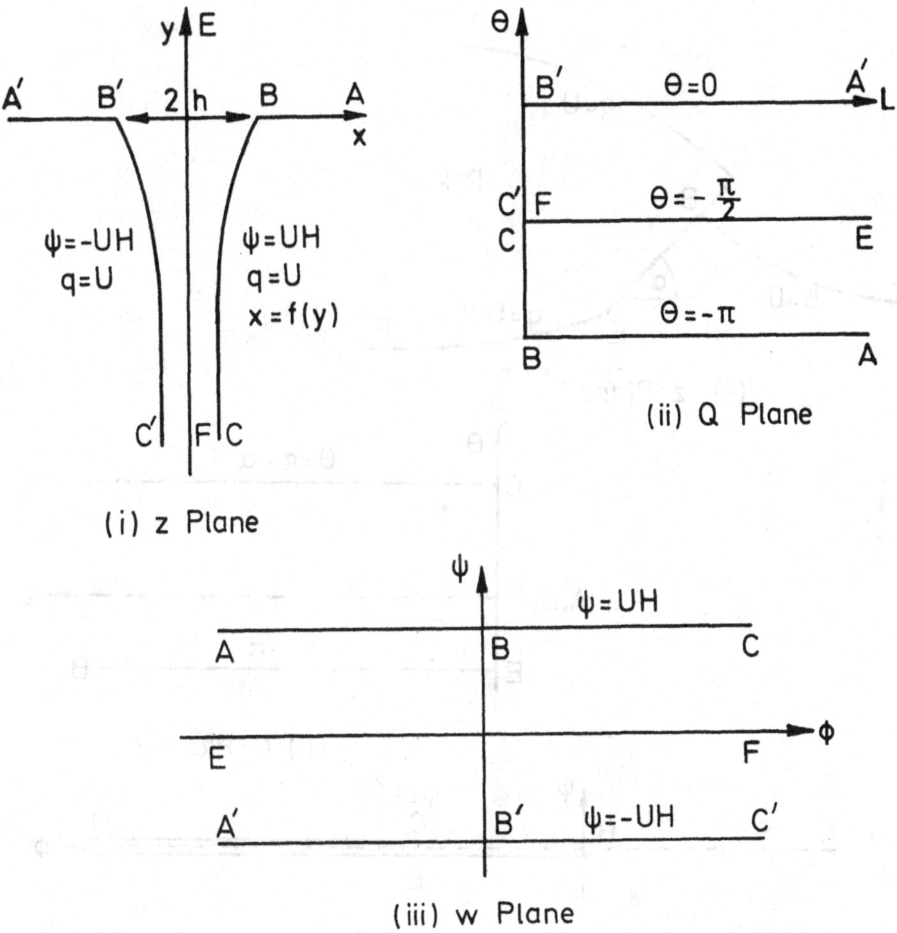

Figure 7.2. Flow of a jet.

and

$$U(\cos\theta - i\sin\theta) = \frac{dw}{dz} = \frac{d\theta}{dz}\frac{d\phi}{d\theta}.$$

Hence $dz/d\theta = (2H/\pi)\tan\theta\,(\cos\theta + i\sin\theta)$ where $z = -h$ when $\theta = 0$, and finally

$$x + iy = z = -h + \frac{2H}{\pi}\left[1 - \cos\theta + i\int_0^\theta \frac{\sin^2\theta'}{\cos\theta'}\,d\theta'\right] \tag{7.3}$$

gives the boundary in parametric form. For large negative values of y, $\theta \to -\pi/2$ and $x \to -h + 2H/\pi = -H$. Thus the contraction ratio $H/h = \pi/(\pi+2)$.

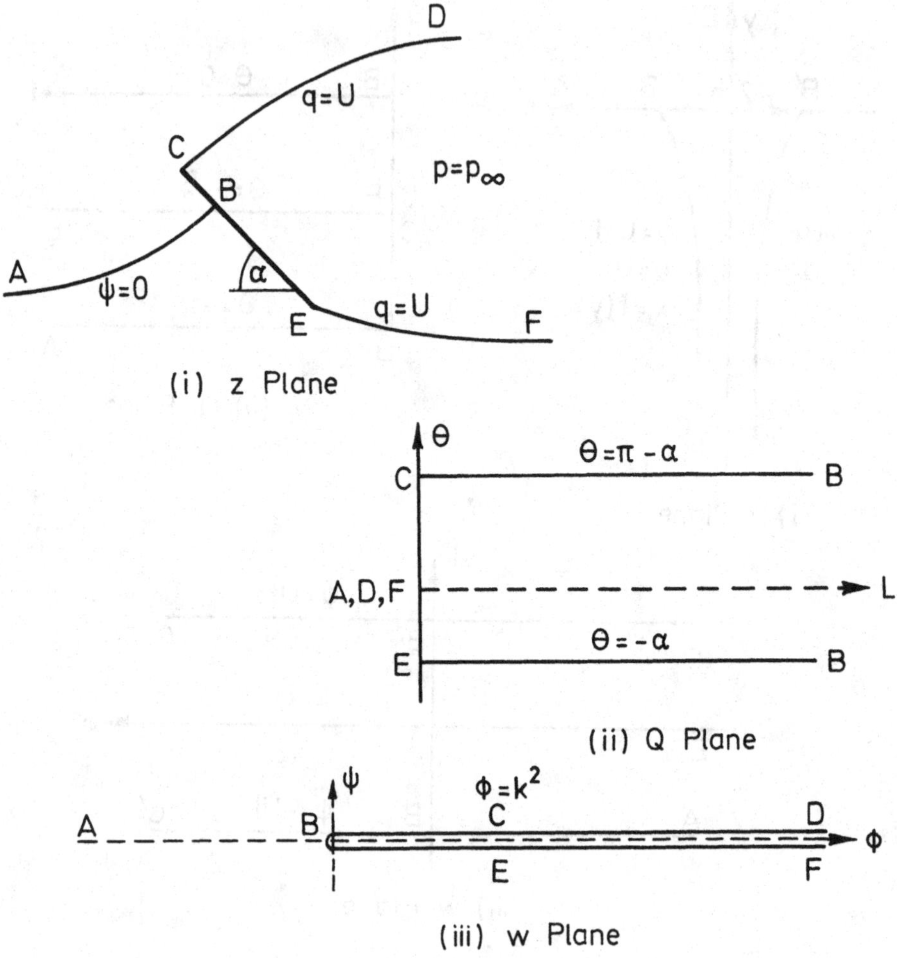

Figure 7.3. Separated flow past a flat plate.

This problem is typical of most jet problems, although other geometries will require a different conformal transformation from the Q to ζ plane.

Example. Separated flow past a flat plate.

A second typical problem concerns the flow past a flat plate at incidence α with a thick wake (i.e., region of constant pressure) forming behind it. The geometrical situation is shown in Figure 7.3 (i), and the regions of fluid in the Q and w planes are shown in Figures 7.3 (ii), (iii). If $\alpha \neq \pi/2$ there is no symmetry, and the dividing streamline shape AB is unknown in both the z and Q planes. The appropriate conformal transformations to the upper half of the ζ plane are

$$\zeta = -\cosh(Q + \alpha i),$$

which takes C into $\zeta = 1$, E into $\zeta = -1$, and A,D,F into $\zeta = -\cos\alpha$, and $\zeta = k(1 + \cos\alpha)/w^{\frac{1}{2}} - \cos\alpha$, where k is real. Uniqueness is assured by fixing the points B, C and D and taking the appropriate branch of $w^{\frac{1}{2}}$. It is interesting to calculate the thrust on the plate, which is given by

$$\int_E^C (p-p_\infty)ds = -\tfrac{1}{2}\rho \int_E^C (U^2-q^2)e^{-i\alpha}d\bar{z} = \tfrac{1}{2}\rho U^2\left(H + \frac{1}{U}\int_E^C e^{-Q-i\alpha}d\bar{w}\right),$$

where the length EC is H. But on EC, $d\bar{w} = +dw = d\phi$, and $Q + \alpha i = \pi i + L$. Hence the non-dimensional thrust is

$$1 - \frac{1}{HU}\int_{-k^2}^{k^2} e^{-L}d\phi = 1 + \frac{2}{HU}\int_0^\infty e^{-L}\frac{d\phi}{dL}\,dL,$$

where $(\cos\alpha + \cosh L)^2 = k^2(1 + \cos\alpha)^2/\phi$ and k is chosen so that

$$H = \int_E^C ds = -\frac{2}{U}\int_0^\infty e^L \frac{d\phi}{dL}\,dL.$$

After some manipulation it may be shown that the thrust on the plate is

$$\rho U^2 H \frac{\pi \sin\alpha}{4+\pi\sin\alpha}. \tag{7.4}$$

2. ATTACHED FLOWS PAST A THIN WING

Consider an incompressible stream U flowing past a wing whose upper and lower surfaces are given by $y = Y_+(x)$ and $y = Y_-(x)$, respectively, for $0 < x < c$. Then if the wing is thin and the flow is attached with the Kutta condition holding at the trailing edge, the boundary condition on the wing may be linearized as in Chapter VI.3. If the velocity potential is given by $U(x+\phi)$, where ϕ is a disturbance potential which vanishes far from the wing, then the boundary conditions reduce to

$$v = \frac{\partial\phi}{\partial y} = Y_\pm'(x) \quad \text{on} \quad y = 0\pm, \quad 0 < x < c, \tag{7.5}$$

as in (6.11) for the compressible case. For $x < 0$ and $x > c$, ϕ and ϕ_y will be continuous, and ϕ will be harmonic everywhere except on

$y = 0$, $0 < x < c$. Note that if trailing vorticity were not neglected, ϕ_x and hence ϕ would be discontinuous on $y = 0$, $x > c$.

Problems which involve finding harmonic functions with discontinuities on a prescribed curve are called Hilbert problems, and there is a general theory using Cauchy integrals, which reduce to singular (principal value) integrals along the curve. (See for example Gakhov [11].) For the thin wing problem, the curve of discontinuity is the slit $y = 0$, $0 < x < c$, and it is convenient to separate the boundary condition (7.5) into the sum of a symmetric and anti-symmetric part. Thus with $\alpha(x) = Y_+'(x) + Y_-'(x)$ and $\beta(x) = Y_+'(x) - Y_-'(x)$, we may consider separately the two problems of flow past a symmetric wing when $\alpha(x) \equiv 0$ and a wing of no thickness when $\beta(x) \equiv 0$, and add the two solutions to solve the general problem.

Consider the Cauchy integral

$$F(z) = \frac{1}{2\pi}\int_0^c \frac{f(t)}{z-t}\,dt,$$

where the path of integration is along the real axis in the complex t plane and $F(z)$ vanishes as $|z| \to \infty$, and for simplicity assume that $f(t)$ is an analytic function of t. Define $F_+(x)$ and $F_-(x)$ as the limiting values of $F(z)$ as z approaches the real axis from above and below respectively. The limits obtained are equivalent to integrals along contours shown in Fig. 7.4, which are below and above the singular point $t = x$ respectively. Thus we can write

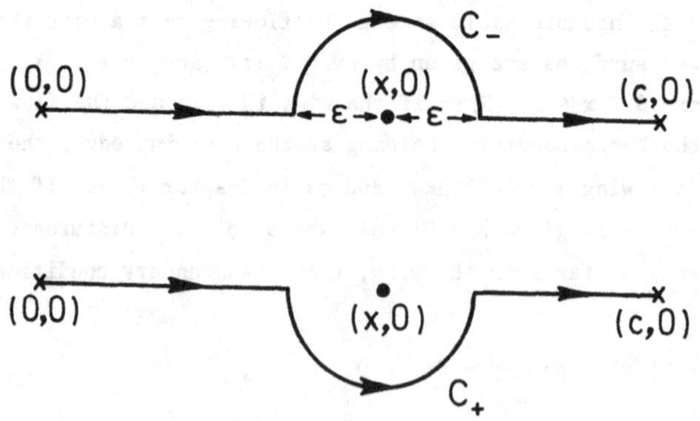

Figure 7.4. Contours in the t-plane.

$$F_{\pm}(x) = \frac{1}{2\pi}\int_{C\pm} \frac{f(t)}{x-t}\,dt,$$

so that by Cauchy's residue theorem

$$F_+ - F_- = \frac{1}{2\pi}\int_{C_+-C_-} \frac{f(t)}{x-t}\,dt = -if(x). \tag{7.6}$$

Also

$$F_+ + F_- = \lim_{\varepsilon,\varepsilon'\to 0} \left\{\frac{1}{\pi}\int_0^{x-\varepsilon} \frac{f(t)}{t-x} + \frac{1}{\pi}\int_{x+\varepsilon'}^{c} \frac{f(t)}{t-x}\,dt\right\},$$

where the path of integration is along the real axis. This limit does not exist for independent parameters ε and ε', but we define the *Cauchy principal value* of the singular integral by the limit obtained when $\varepsilon = \varepsilon'$ and write

$$F_+ + F_- = \frac{1}{\pi}⨍_0^c \frac{f(t)}{x-t}\,dt. \tag{7.7}$$

Equations (7.6) and (7.7) are called the *Plemelj formulae* and may be obtained without assuming that $f(t)$ is analytic, as in Gakhov [11].

The solution to the symmetric thin wing problem may be obtained by direct application of these formulae if we put $F(z) = dw/dz$, where w is the complex disturbance potential, so that

$$\begin{aligned} F_+ + F_- &= (u_+ + u_-) - i(v_+ + v_-),\\ F_+ - F_- &= (u_+ - u_-) - i(v_+ - v_-). \end{aligned} \tag{7.8}$$

Now $v_+ + v_- = 0$ so that $F_+ + F_-$ is real and, from (7.7), f is real. Hence from (7.6), $F_+ - F_-$ is purely imaginary and $u_+ = u_-$, which could have been deduced from the symmetry. Finally

$$F_+ - F_- = -i\beta(x),$$

so that using (7.6) the solution is given by

$$\frac{dw}{dz} = \frac{1}{2\pi}\int_0^c \frac{\beta(t)}{z-t}\,dt,$$

and

$$w = \frac{1}{\pi}\int_0^c Y_+'(t)\log(z-t)\,dt. \tag{7.9}$$

Note that this has the physical interpretation of a distribution of line sources along the wing with density $\beta(t)$.

For the anti-symmetric problem, $v_+ = v_-$, $F_+ - F_-$ is real and $f(t)$ is purely imaginary from (7.6). Thus a distribution of vorticity is needed, and

$$F_+ + F_- = -i\alpha(x) = \frac{1}{\pi}\int_0^c \frac{f(t)}{x-t}\,dt. \tag{7.10}$$

This is a singular integral equation for $f(t)$, for which a solution may be found using the Plemelj formulae. Its solution is not unique, however, and further conditions are needed concerning the behaviour of F near the singular points $z = 0$, $z = c$ and $|z| \to \infty$. For the thin wing problem in which $F = dw/dz$ we require that F vanishes as $|z| \to \infty$; at the trailing edge $z = c$, the Kutta condition requires that F is bounded at $z = c$. At the leading edge the condition is not obvious, and we note that for the flow past a flat plate at incidence (which is an antisymmetric problem if linearized), which was solved in the previous section, dw/dz was unbounded at the leading edge. The difficulty in determining this singularity arises from the fact that for a wing with a smooth leading edge the slope of the wing surface will not be small, as required by the linearized theory, in the vicinity of the leading edge. Thus the approximation is not uniformly valid, and the linearized solution has to be matched to a solution of the full problem valid near the leading edge. Now the wing shape will be approximately parabolic in this region and there will be a continuous flow past it. The appropriate local solution for small enough z is $w = z^{\frac{1}{2}}$, since it is easily verified that this has a parabolic streamline with radius of curvature ε if $\psi = \sqrt{\varepsilon/2}$. Hence the linearized solution for w must behave like $z^{\frac{1}{2}}$ near the leading edge, and we require that $dw/dz \sim z^{-\frac{1}{2}}$. With these conditions we can now proceed to solve for the unique solution of (7.10).

Consider first the problem of constructing a function $G(z)$ such that $G_+ + G_- = 0$ on $y = 0$, $0 < x < c$, with G analytic at all other points. If $Q = \log G$, then

$$Q_+ - Q_- = (2k+1)i\pi, \quad k \text{ any integer},$$

provided some care is taken with zeros of G and consequent branch points of Q. From (7.6)

$$\log G = \frac{1}{2}\int_0^c \frac{(2k+1)}{t-z}\,dt = \frac{2k+1}{2}\log\left(\frac{c-z}{-z}\right),$$

and

$$G(z) = \left(\frac{z-c}{z}\right)^{(2k+1)/2}.$$

In addition to these solutions we can multiply G by any function analytic everywhere except at $z = 0$ and c, namely z^n or $(z-c)^n$ where n is any integer, so that possible functions G are $(z-c)^{n_1+\frac{1}{2}}z^{n_2+\frac{1}{2}}$ where n_1 and n_2 are integers.

Define a new function $H = GF$ so that

$$H_+ - H_- = G_+(F_+ + F_-) = -i\alpha(x)G_+(x),$$

$$H_+ + H_- = G_+(F_+ - F_-) = -if(x)G_+(x).$$

Then, since H satisfies the Plemelj formulae (7.6) and (7.7),

$$f(x)G_+(x) = \frac{i}{\pi} \int_0^c \frac{\alpha(t)G_+(t)}{x-t}\, dt \tag{7.11}$$

gives the appropriate distribution of vorticity. The complex potential may, however, be obtained directly without recourse to substitution of (7.11) in the integral for $F(z)$, since

$$H(z) = \frac{1}{2\pi} \int_0^c \frac{\alpha(t)G_+(t)}{z-t}\, dt.$$

Thus

$$\frac{dw}{dz} = F(z) = \frac{1}{2\pi G(z)} \int_0^c \frac{\alpha(t)G_+(t)}{z-t}\, dt, \tag{7.12}$$

and this may be integrated with respect to z to obtain w. Since dw/dz vanishes as $|z| \to \infty$, G must not vanish as $|z| \to \infty$, and $n_1 + n_2 \geq -1$. From the Kutta condition at the trailing edge $n_1 \leq -1$, and from the leading edge matching condition $n_2 = 0$, so that

$$G(z) = \left(\frac{z}{z-c}\right)^{\frac{1}{2}}$$

is the unique function for G (up to an arbitrary multiplicative constant). Thus, from (7.12), the solution to the antisymmetric thin wing problem is given by

$$\frac{dw}{dz} = \frac{i}{2\pi} \left(\frac{z-c}{z}\right)^{\frac{1}{2}} \int_0^c \left(\frac{t}{c-t}\right)^{\frac{1}{2}} \frac{\alpha(t)}{t-z}\, dt. \tag{7.13}$$

For large values of z,

$$\frac{dw}{dz} \sim -\frac{i}{2\pi z} \int_0^c \left(\frac{t}{c-t}\right)^{\frac{1}{2}} \alpha(t)dt,$$

so that the circulation Γ is

$$-U\int_0^c \left(\frac{t}{c-t}\right)^{\frac{1}{2}} \alpha(t)dt.$$

In the case of the flat plate at incidence α_0, $\alpha(t) = -2\alpha_0$ and $\Gamma = \pi\alpha_0 Uc$. This is equivalent to the exact result obtained in Section VII.1 when $c = 4$ and α_0 is small.

3. SEPARATED FLOWS PAST A THIN WING

If the flow separates at the point $x = b$ on the upper surface to form a thick wake as in Fig. 7.5 (i), then from Bernoulli's equation the boundary condition on $y = 0+$, for $b < x < 1$, is $u_+ = 0$. For the anti-symmetric problem of a wing of zero thickness,† $u = 0$ on $y = 0$ for $x > c$ so that the boundary conditions may be written

$$y = 0+,\quad v_+ = \tfrac{1}{2}\alpha(x),\quad 0 < x < b \quad\text{and}\quad u_+ = 0,\quad x > b;$$

$$y = 0-,\quad v_- = \tfrac{1}{2}\alpha(x),\quad 0 < x < c \quad\text{and}\quad u_- = 0,\quad x > c.$$

This is an example of a Riemann-Hilbert problem in which a linear combination of the real and imaginary parts of an analytic function is given on a closed curve, which in this case consists of the upper and lower sides of the real axis with $x > 0$, and the circle at infinity. We solve such problems by transforming the region enclosed by the curve to the upper half plane, and for this example the transformation is clearly $\zeta = z^{\frac{1}{2}}$ as in Fig. 7.5 (ii). Consider the analytic function $G(\zeta) = dw/d\zeta = 2\zeta(u-iv)$, defined in $\eta > 0$ where $\zeta = \xi+i\eta$. Define $G(\xi)$ in $\eta < 0$ by $G(\zeta) = \overline{G(\bar{\zeta})}$. Then G is an analytic function of ζ, except on the real axis $\eta = 0$, and has to be bounded for large $|\zeta|$ since $|u-iv| \to 0$ as $|z| \to \infty$. On the real axis,

$$G_+ = -2iv\xi,\quad G_- = 2iv\xi \quad\text{so that}\quad G_++G_- = 0 \quad\text{for}\quad \xi < -\sqrt{c},\quad \xi > \sqrt{b};$$

$$G_+ = 2\xi u-i\xi\tilde{\alpha}(\xi),\quad G_- = 2\xi u + i\xi\tilde{\alpha}(\xi) \quad\text{so that}\quad G_+-G_- = -2i\xi\tilde{\alpha}(\xi)$$

$$\text{for}\quad -\sqrt{c} < \xi < \sqrt{b},\ \text{where}\quad \tilde{\alpha}(\xi) = \alpha(x(\xi)).$$

We now look for a function $H(\zeta)$ satisfying $H_++H_- = 0$ on the real axis of the ζ plane exterior to the slit $(-\sqrt{c}, \sqrt{b})$. Then

$$(GH)_+ - (GH)_- = 0,\quad \xi < -\sqrt{c},\quad \xi > \sqrt{b};$$

$$= -2i\xi\tilde{\alpha}(\xi)H(\xi),\quad -\sqrt{c} < \xi < \sqrt{b}.$$

†We are grateful to Professor D. A. Spence for the formulation and solution of this problem.

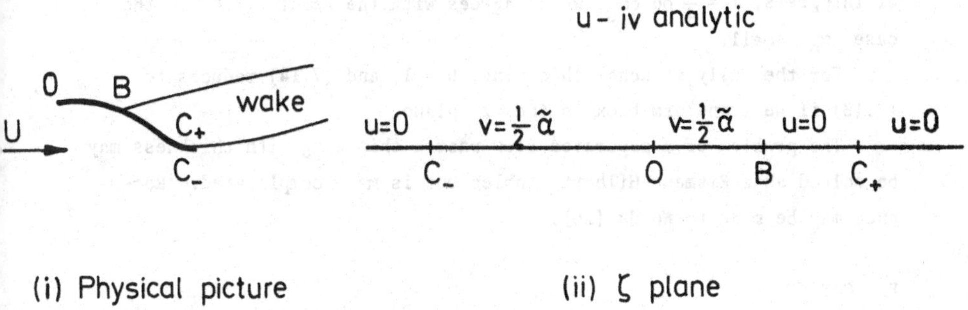

Figure 7.5. Separated flow past a thin wing.

From the Plemelj formula (7.7),

$$G(\zeta)H(\zeta) = \frac{1}{\pi}\int_{-\sqrt{c}}^{\sqrt{b}} \frac{t\tilde{\alpha}(t)H(t)}{\zeta-t}\,dt,$$

and

$$\frac{dw}{dz} = \frac{1}{2\pi\zeta H(\zeta)}\int_{-\sqrt{c}}^{\sqrt{b}} \frac{t\tilde{\alpha}(t)H(t)}{\zeta-t}\,dt. \tag{7.14}$$

Possible functions $H(\zeta)$ are $(\zeta+\sqrt{c})^{n_1+\frac{1}{2}}(\zeta-\sqrt{b})^{n_2+\frac{1}{2}}$ where n_1 and n_2 are integers. For dw/dz to vanish as $|\zeta| \to \infty$, $n_1+n_2 \geq -2$; for dw/dz to be bounded at the trailing edge and separation point, $n_1 \leq -1$, $n_2 \leq -1$. Hence $n_1 = n_2 = -1$ and $H = \{(\zeta+\sqrt{c})(\zeta-\sqrt{b})\}^{-\frac{1}{2}}$. Note that dw/dz is unbounded at the leading edge $\zeta = 0 = z$ and has a square root singularity as in the attached flow problem of the previous section.

We can now calculate the lift force Y exerted by the stream on the wing from the expression $\rho U^2 \int_0^c (u_+-u_-)dx$, which may be rewritten as $2\rho U^2 \int_{-\sqrt{c}}^{\sqrt{b}} u\xi d\xi$ in the ζ plane. To evaluate u we note that

$$G_+ + G_- = \frac{2}{\pi H(\xi)}\int_{-\sqrt{c}}^{\sqrt{b}} \frac{H(t)t\tilde{\alpha}(t)}{\xi-t}\,dt \quad \text{for} \quad -\sqrt{c} < \xi < \sqrt{b},$$

so that

$$u = \frac{(\xi+\sqrt{c})^{\frac{1}{2}}(\xi-\sqrt{b})^{\frac{1}{2}}}{2\pi\xi}\int_{-\sqrt{c}}^{\sqrt{b}} \frac{t\tilde{\alpha}(t)\,dt}{(t+\sqrt{c})^{\frac{1}{2}}(t-\sqrt{b})^{\frac{1}{2}}(\xi-t)}, \tag{7.15}$$

and the lift force can be calculated. In the case of a fully separated flat plate aerofoil, $b = 0$ and $\tilde{\alpha}(t) = -2\alpha_0$. After some manipulation of integrals, $Y = \frac{\pi}{4}\rho U^2 c\alpha_0$, which agrees with the result (7.4) in the case α_0 small.

For the fully attached thin wing, $b = 1$, and (7.14) reduces to (7.13) if we transform back to the z plane.

The problem of a separated flow past a thin wing with thickness may be solved as a Riemann-Hilbert problem but is more complicated. Reference may be made to Woods [26].

EXERCISES

1. A steady two-dimensional jet of incompressible fluid, speed U and thickness h, strikes an infinite plane wall normally. Show that in the Q plane the fluid occupies the region $0 \leq L < \infty$, $-\pi/2 < \theta < \pi/2$, and in the w plane $-Uh/2 < \psi < Uh/2$, $-\infty < \phi < \infty$. Verify that the region in the Q plane is mapped onto that in the w plane by the conformal transformation

$$w = \frac{hU}{\pi}\log\tanh Q.$$

Obtain parametric equations (in terms of θ) for the jet boundary.

2. (i) A two dimensional aerofoil consists of a circular arc whose chord length (distance between end points) is 4 and whose angle subtended by this chord is 2β. It is placed in a stream U at incidence α to the chord, and there is a circulation Γ. Obtain an expression for the velocity potential $w(\zeta)$ for the flow by first verifying that the conformal transformation

$$\frac{\zeta-2}{\zeta+2} = \left(\frac{z-1}{z+1}\right)^2$$

takes the exterior of the arc in the ζ plane into the exterior of the circle $|z - i\cot\beta| = \operatorname{cosec}\beta$, where $0 < \beta < \pi/2$. Show that the velocity will be bounded at the trailing edge if

$$\Gamma = 4\pi U\,\frac{\cos(\beta-\alpha)}{\sin\beta}\,.$$

(ii) If $\beta = \frac{1}{2}\pi-\varepsilon$ and both ε and α are small so that the circular arc is a thin wing, show that the surface of the wing is given by $Y_\pm(x) = \varepsilon/2(4-x^2)-\alpha x$, where now the x axis is in the free stream direction. Hence obtain an expression for the velocity potential and show that $\Gamma = 4\pi U(\alpha+\varepsilon)$, in agreement with the result in (i).

3. Derive the integral relations

$$w_+(x) + w_-(x) = \frac{1}{i\pi} \fint_C \frac{w_+(t) - w_-(t)}{t-x}\,dt$$

connecting the perturbation velocities in a thin wing problem, where $w = u - iv$ and $\pm$ denotes values on opposite sides of C.

A thin diamond shaped obstacle, with diagonals of lengths $2a$, $2\alpha a$ ($\alpha \ll 1$) is placed with two of its opposite vertices at points $z = \pm a$ in a uniform stream of incompressible fluid, having velocity U at infinity parallel to the real axis. Show that in the absence of circulation the pressure at a point x on the surface is

$$p_\infty + \rho \frac{u^2\alpha}{\pi} \log\left(\frac{x^2}{a^2 - x^2}\right)$$

according to thin aerofoil theory. Show that this result corresponds near $x = a$ to a simple solution of Laplace's equation near a sharp corner.

4. Show that the singular integral equation

$$a(x)f(x) = \frac{\lambda}{\pi i} \fint_C \frac{f(t)dt}{t-x} + g(x), \quad (x \in C),$$

where C is a smooth closed contour in the complex plane and $\fint$ denotes the Cauchy principal value, can be reduced to a Hilbert problem for the boundary values of the function

$$F(z) = \frac{1}{2\pi i} \int_C \frac{f(t)dt}{t-z}$$

on C.

If $a(x) = x$, and the points $z = \pm\lambda$ lie respectively outside and inside C, show that in the case $g \equiv 0$, the Hilbert problem is satisfied by the function defined as

$$F(z) = \begin{cases} (z-\lambda)^{-1}, & z \text{ inside } C, \\ (z+\lambda)^{-1}, & z \text{ outside } C. \end{cases}$$

Deduce that when $g(x) \neq 0$, the solution of the integral equation is

$$f(x) = (x^2 - \lambda^2)^{-1}\left[\frac{\lambda}{\pi i} \fint \frac{g(t)dt}{t-x} + xg(x) + k\right],$$

where k is an arbitrary constant.

Appendix
Hyperbolic Equations With Two Independent Variables

1. SECOND ORDER SCALAR EQUATIONS

Consider the general quasi-linear equation

$$a\phi_{xx} + 2b\phi_{xy} + c\phi_{yy} = f(x,y,\phi,\phi_x,\phi_y), \tag{A.1}$$

where a, b and c are functions of x, y, ϕ, ϕ_x and ϕ_y, with Cauchy boundary data on an open curve; that is,

$$x = x_0(s), \quad y = y_0(s), \quad \phi_x = \phi_{ox}(s), \quad \phi_y = \phi_{oy}(s), \tag{A.2}$$

where s is a parameter. Note that by integration along C, $\phi = \phi_0(s)$ is prescribed except for a constant. If this constant is specified then a necessary condition for (A.1) and (A.2) to define a well-posed boundary value problem for ϕ is that by differentiating along Γ the derivatives of all orders may be calculated. Differentiating once

$$\begin{aligned} \phi'_{0x} &= x'_0\phi_{xx} + y'_0\phi_{xy}, \\ \phi'_{0y} &= x'_0\phi_{xy} + y'_0\phi_{yy}, \end{aligned} \tag{A.3}$$

so that a necessary condition is that there is a unique solution of equations (A.1) and (A.3) for ϕ_{xx}, ϕ_{xy} and ϕ_{yy}. This implies that

$$ay_0'^2 - 2bx_0'y_0' + cx_0'^2 \neq 0, \tag{A.4}$$

and it may be shown that this condition is also sufficient to ensure the existence of derivatives of all orders if the coefficients and boundary values are infinitely differentiable.

We define a *characteristic* to be a curve in the x,y plane along which Cauchy data does not uniquely define the second derivatives of ϕ.

Hence given $x(t)$, $y(t)$, $\phi_x(t)$, $\phi_y(t)$ on a characteristic, we have

$$\dot{\phi}_x = \dot{x}\phi_{xx} + \dot{y}\phi_{xy},$$
$$\dot{\phi}_y = \dot{x}\phi_{xy} + \dot{y}\phi_{yy}, \tag{A.5}$$

which together with (A.1) must lead to a non-unique solution for the second derivatives ϕ_{xx}, ϕ_{xy} and ϕ_{yy}. Thus equations (A.1) and (A.5) are linearly dependent, and this implies that

$$\begin{vmatrix} a & 2b & c \\ \dot{x} & \dot{y} & 0 \\ 0 & \dot{x} & \dot{y} \end{vmatrix} = 0 = \begin{vmatrix} a & f & c \\ \dot{x} & \dot{\phi}_x & 0 \\ 0 & \dot{\phi}_y & \dot{y} \end{vmatrix},$$

which reduces to

$$a\dot{y}^2 - 2b\dot{x}\dot{y} + c\dot{x}^2 = 0 \tag{A.6}$$

and

$$a\dot{y}\dot{\phi}_x + c\dot{x}\dot{\phi}_y = f\dot{x}\dot{y}. \tag{A.7}$$

In the semi-linear case, when a, b and c are functions of x and y only, (A.6) determines the characteristics, which are real if $b^2 \geq ac$, with two real characteristic directions at a point if $b^2 > ac$. In the quasi-linear case when a,b,c also depend on ϕ, ϕ_x, ϕ_y, this is still a necessary condition for two real characteristic directions to exist at a point, but the directions cannot in general be evaluated without a knowledge of the solution ϕ. If two real characteristic directions exist at a point the equation is said to be locally *hyperbolic*, and if this is true at all points in the domain D on one side of the open boundary defined by (A.2), then the equation is hyperbolic in D and the Cauchy-Kowaleski theorem gives the existence of a solution ϕ in some neighborhood of the boundary. (See Garabedian [12].) Note that these conditions require that the boundary curve does not touch a characteristic of either family.

For a hyperbolic equation, the characteristics also define the "region of influence" of the initial data. Equation (A.7) shows that there is a relation between the derivatives of ϕ_x and ϕ_y along a characteristic, and it can be shown (see Courant & Hilbert, Vol. II [8]) that the solution at P depends on the initial data between A and B, where AP and BP are the positive and negative characteristics through P as shown in Figure A.1. Conversely, Cauchy data given on AB defines a unique solution in the region APB bounded by the characteristics through A

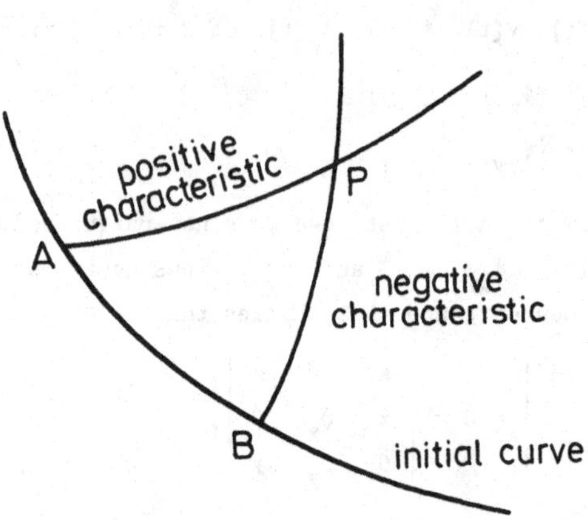

Figure A.1

and B. For two dimensional supersonic flow, the characteristics are the Mach lines and the "region of influence" of a body may be identified with the part of the fluid which is not in the "zone of silence" described on p. 64.

In a number of examples it is possible to integrate (A.7) explicitly along a characteristic of either family and define two functions, called *Riemann invariants*, which are constant along characteristics of the appropriate family. The existence of these invariants allows solutions to be constructed for boundary value problems of a special class, called simple wave solutions, of which the two dimensional steady Prandtl-Meyer flow of a gas described in Chapter IV is an example. If we apply the theory to equation (4.19), then the characteristic directions are defined by

$$(a^2-u^2)\dot{y}^2 + 2uv\,\dot{x}\dot{y} + (a^2-v^2)\dot{x}^2 = 0,$$

which reduces to equation (4.24), namely

$$\frac{dy}{dx} = \frac{-uv \pm a[u^2+v^2-a^2]^{\frac{1}{2}}}{a^2-u^2}.$$

The characteristics exist if $u^2+v^2 > a^2$, that is if the flow is supersonic. On a characteristic, (A.7) gives

$$(a^2-u^2)\dot{y}\dot{u} + (a^2-v^2)\dot{x}\dot{v} = 0,$$

which is relation (4.25).

The Riemann invariants are complicated functions in terms of u and v, but rewritten in terms of μ and θ, where $\sin\mu = a/(u^2+v^2)^{\frac{1}{2}}$ and $\tan\theta = v/u$, they are $\theta \pm f(\mu)$, as in equations (4.27) and (4.28). In this example the characteristics are physically identified with the Mach lines of the flow.

2. FIRST ORDER SYSTEMS OF EQUATIONS

Consider the first order system

$$Au_x + Bu_y = c, \tag{A.8}$$

where $u \in \mathbb{R}^n$ and A and B are $n \times n$ matrices. The system is quasi-linear if the entries and components of A, B and c are functions of x, y and u, and linear if they are functions of x and y alone.

If Cauchy boundary data $u = u_0(s)$ on $x = x_0(s)$, $y = y_0(s)$ is given, then a necessary condition for a well posed problem is that the partial derivatives of u may all be uniquely determined on the boundary. Hence, from (A.8) evaluated on the boundary and the relation

$$u_0' = x_0' u_x + y_0' u_y,$$

we require a unique solution of $2n$ simultaneous equations for the $2n$ components of u_x and u_y. This implies that the $2n \times 2n$ determinant of the coefficients of u_x and u_y is non-zero, and the condition can be simplified to $|Bx_0' - Ay_0'| \neq 0$. This corresponds to the condition (A.4) of the previous section and in the case $n = 2$ will be a quadratic function of x_0' and y_0' as in (A.4). Indeed if the coefficients in (A.1) are independent of ϕ, it may easily be written as a first order system by the substitution $u_1 = \phi_x$, $u_2 = \phi_y$.

As in the previous section we may define characteristics as curves in the x,y plane along which Cauchy data does not uniquely define the first derivatives of u. However, a more direct approach, which is easily shown to be equivalent, is to define a characteristic $(x(t),y(t))$ by

$$|B\dot{x} - A\dot{y}| = 0. \tag{A.9}$$

A system is hyperbolic at a point if there exist n characteristic directions, that is, if (A.9) has n real roots for $\lambda = \dot{y}/\dot{x}$. For such an eigenvalue λ we can define a left eigenvector z such that $z^T(B-\lambda A) = 0$. Then, using (A.8),

$$z^T A(u_x + \lambda u_y) = z^T c,$$

and hence along a characteristic

$$z^T A \frac{du}{dx} = z^T c \tag{A.10}$$

is an ordinary differential equation for u. If this equation is integrable, then the resultant function which is constant along the characteristic is called a Riemann invariant.

A simple example is that of the shallow water theory equations described in Chapter III.1, where variables t and x replace x and y. For equations (3.3) and (3.4),

$$u = \begin{pmatrix} u \\ \eta \end{pmatrix}, \quad A = \begin{pmatrix} 0 & 1 \\ 1 & 0 \end{pmatrix}, \quad B = \begin{pmatrix} \eta & u \\ u & g \end{pmatrix}, \quad \text{and} \quad c = 0.$$

The roots of $\begin{vmatrix} \eta & u-\lambda \\ u-\lambda & g \end{vmatrix} = 0$ are $\lambda = u \pm \sqrt{g\eta}$ so that the system is hyperbolic for $\eta > 0$. The left eigenvector $z^T = (z_1, z_2)$ satisfies $\eta z_1 \mp z_2\sqrt{g\eta} = 0$, so that from (A.10)

$$0 = z^T A \frac{du}{dx} = (\eta, \pm\sqrt{g\eta})\frac{du}{dx} = \eta \frac{du}{dx} \pm \sqrt{g\eta}\, \frac{d\eta}{dx} .$$

This is integrable and gives Riemann invariants $u \pm 2\sqrt{g\eta}$, as obtained by elementary manipulation in (3.5).

A second example is the one dimensional flow of a gas described in Chapter IV.3 and governed by equations (1.6) and (1.8). In this case $u = \begin{pmatrix} u \\ \rho \end{pmatrix}$, $A = \begin{pmatrix} 0 & 1 \\ 1 & 0 \end{pmatrix}$ and $B = \begin{bmatrix} \rho & u \\ u & a^2/\rho \end{bmatrix}$. An example with $n = 3$ is the two dimensional steady flow of an isentropic gas discussed in Chapter IV.4, which is described in the previous section in the homentropic case, when it may be reduced to a second order scalar equation. In the general case the equations are (4.16), (4.17) and (4.18), and

$$u = \begin{bmatrix} u \\ v \\ \rho \end{bmatrix}, \quad A = \begin{bmatrix} \rho & 0 & u \\ u & 0 & a^2/\rho \\ 0 & u & 0 \end{bmatrix} \quad \text{and} \quad B = \begin{bmatrix} 0 & \rho & v \\ v & 0 & 0 \\ 0 & v & a^2/\rho \end{bmatrix}.$$

It is easily verified that in the supersonic case there are three families of characteristics, the two Mach lines and the streamlines. However, only one Riemann invariant exists if the flow is not homentropic, and that is along the streamlines where

$$(u^2+v^2) + \int \frac{a^2 d\rho}{\rho} = \text{const.},$$

which is Bernoulli's equation.

One property of the characteristics of (A.1) and (A.8) is that they are curves across which discontinuities in the second derivatives of ϕ or the first derivatives of u may occur. Thus a physical interpretation of a characteristic in terms of gas dynamics is that of a weak shock wave. For a linear second order equation or system the characteristics are also curves across which discontinuities in $\nabla\phi$ or u may occur. Thus in the linearized theory of Chapter VI, shock waves lie along characteristics. However, for non-linear problems shock waves do not lie along characteristics, and Riemann invariants are discontinuous at shock waves.

References

[1] Batchelor, G. K. An Introduction to Fluid Dynamics. CUP, 1967.

[2] Becker, E. Gas Dynamics. Academic Press, 1968.

[3] Birkhoff, G. and Zarantonello, E. H. Jets, Wakes & Cavities. Academic Press, 1957.

[4] Carrier, G. F. and Pearson, C. E. Ordinary Differential Equations. Blaisdell, 1968.

[5] Chester, C. R. Techniques in Partial Differential Equations. McGraw-Hill, 1971.

[6] Cole, J. Perturbation Methods in Applied Mathematics. Blaisdell, 1968.

[7] Copson, E. T. Asymptotic Expansions. CUP, 1967.

[8] Courant, R. and Hilbert, D. Methods of Mathematical Physics, Vol. II. Interscience, 1962.

[9] Erdelyi, A. Higher Transcendental Functions, Vol. II. McGraw-Hill, 1955.

[10] Erdelyi, A. Tables of Integral Transforms, Vol. I. McGraw-Hill, 1954.

[11] Gakhov, S. D. Boundary Value Problems. Pergamon, 1966.

[12] Garabedian, P. R. Partial Differential Equations. Wiley, 1963.

[13] Greenspan, H. The Theory of Rotating Fluids. CUP, 1968.

[14] Hayes, W. D. and Probstein, R. F. Hypersonic Flow Theory. Academic Press, 1966.

[15] Howarth, L. (Ed.) Modern Developments in Fluid Dynamics, High Speed Flow. OUP, 1953, Vol. I.

[16] Landau, L. D. and Lifschitz, E. M. Fluid Mechanics. Pergamon, 1959.

[17] Liepmann, H. W. and Roshko, A. Elements of Gas Dynamics. Wiley, 1957.

[18] Milne Thomson, L. M. Theoretical Hydrodynamics. Macmillan, 1962.

[19] Murray, J. D. Asymptotic Analysis. OUP, 1974.

[20] Officer, C. B. Introduction to the Theory of Sound Transmissions. McGraw Hill, 1958.

[21] Rosenhead, L. (Ed.) Laminar Boundary Layers. OUP, 1963.

[22] Schiffer, M. Article F, Handbuch der Physik IX. Springer-Verlag, 1960.

[23] Schlichting, H. Boundary Layer Theory. McGraw-Hill, 1955.

[24] Stoker, J. J. Water Waves. Interscience, 1957.

[25] Van Dyke, M. D. Perturbation Methods in Fluid Mechanics. Parabolic Press, 1975.

[26] Woods, L. C. The Theory of Subsonic Plane Flow. CUP, 1961.

[27] Yih, C. S. Dynamics of Nonhomogeneous Fluids. Macmillan, 1965.

Index

Applied Mathematical Sciences

1. John: **Partial Differential Equations**, 4th ed. (cloth)
2. Sirovich: **Techniques of Asymptotic Analysis.**
3. Hale: **Theory of Functional Differential Equations**, 2nd ed. (cloth)
4. Percus: **Combinatorial Methods.**
5. von Mises/Friedrichs: **Fluid Dynamics.**
6. Freiberger/Grenander: **A Short Course in Computational Probability and Statistics.**
7. Pipkin: **Lectures on Viscoelasticity Theory.**
8. Giacaglia: **Perturbation Methods in Non-Linear Systems.**
9. Friedrichs: **Spectral Theory of Operators in Hilbert Space.**
10. Stroud: **Numerical Quadrature and Solution of Ordinary Differential Equations.**
11. Wolovich: **Linear Multivariable Systems.**
12. Berkovitz: **Optimal Control Theory.**
13. Bluman/Cole: **Similarity Methods for Differential Equations.**
14. Yoshizawa: **Stability Theory and the Existence of Periodic Solutions and Almost Periodic Solutions.**
15. Braun: **Differential Equations and Their Applications**, 3rd ed. (cloth)
16. Lefschetz: **Applications of Algebraic Topology.**
17. Collatz/Wetterling: **Optimization Problems.**
18. Grenander: **Pattern Synthesis: Lectures in Pattern Theory, Vol I.**
19. Marsden/McCracken: **The Hopf Bifurcation and its Applications.**
20. Driver: **Ordinary and Delay Differential Equations.**
21. Courant/Friedrichs: **Supersonic Flow and Shock Waves.** (cloth)
22. Rouche/Habets/Laloy: **Stability Theory by Liapunov's Direct Method.**
23. Lamperti: **Stochastic Processes: A Survey of the Mathematical Theory.**
24. Grenander: **Pattern Analysis: Lectures in Pattern Theory, Vol. II.**
25. Davies: **Integral Transforms and Their Applications.**
26. Kushner/Clark: **Stochastic Approximation Methods for Constrained and Unconstrained Systems.**
27. de Boor: **A Practical Guide to Splines.**
28. Keilson: **Markov Chain Models—Rarity and Exponentiality.**
29. de Veubeke: **A Course in Elasticity.**
30. Sniatycki: **Geometric Quantization and Quantum Mechanics.**
31. Reid: **Sturmian Theory for Ordinary Differential Equations.**
32. Meis/Markowitz: **Numerical Solution of Partial Differential Equations.**
33. Grenander: **Regular Structures: Lectures in Pattern Theory, Vol. III.**
34. Kevorkian/Cole: **Perturbation Methods in Applied Mathematics.** (cloth)
35. Carr: **Applications of Centre Manifold Theory.**
36. Bengtsson/Ghil/Källén: **Dynamic Meterology: Data Assimilation Methods.**
37. Saperstone: **Semidynamical Systems in Infinite Dimensional Spaces.**
38. Lichtenberg/Lieberman: **Regular and Stochastic Motion.** (cloth)
39. Piccinini/Stampacchia/Vidossich: **Ordinary Differential Equations in R^n.**
40. Naylor/Sell: **Linear Operator Theory in Engineering and Science.** (cloth)
41. Sparrow: **The Lorenz Equations: Bifurcations, Chaos, and Strange Attractors.**
42. Guckenheimer: **Nonlinear Oscillations, Dynamical Systems and Bifurcations of Vector Fields.**
43. Ockendon/Tayler: **Inviscid Fluid Flows.**